PHARMAKAKAWA

PHARMAKAKAWA

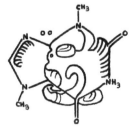

Theobroma Cacao in Medicine and Ceremony

Marcos Patchett

AEON

First published in 2024 by
Aeon Books

British Library Cataloguing in Publication Data

A C.I.P. for this book is available from the British Library

ISBN-13: 978-1-80152-139-0

Typeset by Medlar Publishing Solutions Pvt Ltd, India

www.aeonbooks.co.uk

*For Liana, my parents, and my family (including in-laws),
who probably won't read this book;*

and for Mel, Lynne, Wendy, Vagelis, and Florian, who might.

CONTENTS

PREFACE

Following the publication of *The Secret Life of Chocolate* in 2020, there was a very positive response to the book from a small but enthusiastic cadre of chocolate lovers, Cacao ceremony enthusiasts, and fellow herbalists. Despite being published in the week of COVID lockdown in the UK and removed from Amazon for the first three months of its release, it gradually began to sell slowly but surely. A 700-plus page hardback fourteen years in the making, *The Secret Life of Chocolate* isn't a budget offering; so perhaps unsurprisingly, in addition to receiving many kind and positive comments, I heard "I'd love to buy your book, but it's just so expensive!", or "I'm waiting for my birthday/Christmas/the rapture as I just can't afford it at the moment", and variations on that theme.

So, when my publisher suggested an abridged version of *The Secret Life of Chocolate* focusing on the medicinal and ceremonial uses of Cacao, it didn't take me long to decide that this was a good idea. I could take the opportunity to focus on the main areas of interest for a significant part of the book's audience and for myself as a practicing medical herbalist. *Pharmakakawa* is the result. It doesn't contain all the information, nuance, detail or breadth of information in *The Secret Life of Chocolate*—over 200,000 words don't easily compress into under 20,000 words, after all, but it distils a few core concepts as to how *Theobroma cacao* may be served to maximise its therapeutic potential. This book is a brief guide to what I believe are radical reasons for Cacao's high esteem in Mesoamerica: its sustaining capacity in personal and community health, and its potential to facilitate desired changes in individuals and societies who interact with it mindfully.

DISCLAIMER

The information in this book should be used responsibly. The author does not endorse the use of any illegal substances. Any attempt to replicate the formulas in this book or to utilise the information contained herein is undertaken at the discretion of the reader, and is their responsibility; the author accepts no liability for any harm arising from medicinal or culinary uses of chocolate or other substances described in this book. All medicinal recommendations should be checked with your qualified medical herbalist, physician, or accredited health professional.

I've endeavoured to present the historical, scientific, anthropological, and philosophical information in this book accurately, but I have been known to make mistakes; any errors, misapprehensions, or non-sequiturs are unlikely to be the fault of cited authors.

CHAPTER ONE

Vessels

"**vessel**: noun. 1a: a container (such as a cask, bottle, kettle, cup, or bowl) for holding something. 1b: a person into whom some quality (such as grace) is infused—"a child of light, a true *vessel* of the Lord" —H. J. Laski. 2: a watercraft bigger than a rowboat […] 3a: a tube or canal (such as an artery) in which a body fluid is contained and conveyed or circulated. 3b: a conducting tube in the xylem of a vascular plant formed by the fusion and loss of end walls of a series of cells."

(Merriam-Webster, 2023)

Theobroma cacao is a compact understorey tree in the mallow family (*Malvaceae*) which originated in the Americas but diffused across the globe in colonial ships, and now engirdles the world in two narrow bands of latitude around twenty degrees north and south of the equator. Cacao thrives in the humidity, shade, and humus-enriched soil provided by the shelter of taller trees. The nutritious leaf litter of the tropical forest floor sustains various microbes, some of which are symbionts: lateral 'feeder roots' from the tree spread out and collude with fungal mycelia, a *mycorrhizal* system in which each organism supplies the other with sustenance akin to the human microbiome or 'gut flora'.

Cacao trees look unusual, flowering and fruiting directly from the trunk. Clusters of tiny five-petalled star-shaped flowers bloom at the start of the rainy season, and are pollinated by 'chocolate midges' (of various families) and stingless bees (*Meliponinae*). The fertilised flowers burgeon into green, thick-skinned ovoid pods that flush red, yellow, purple, or brown as they mature. Their inch-thick rind holds an average of thirty to forty-five "beans" or seeds, packed in a pentagonal arrangement in cross-section, and embedded in a matrix of sweet, white, mucilaginous fruit pulp. The trees begin to fruit at two or three years old, attaining maximum yields from six or seven years of age. They require high annual rainfall[1] and prefer temperatures above 27°C (81°F) but won't tolerate less than 16°C (64°F) or dry weather for any length of time, nor will they grow more than 700m above sea level.[2] They are very susceptible to diseases and pests, and easily damaged by high winds.[3] Part of the reason for this sensitivity may be centuries of human agricultural interaction, with selective breeding to increase crop yields.

There are ten genetic "clusters" of Cacao,[4] but only three principal varieties: *criollo*, the original Mesoamerican (pre-colonial Central American) cultivar, accounting for only 5% of the world's Cacao crop;[5] the hardier, more fertile *forastero*, from Amazonia in South America; and *mestizo* or *trinitario*, so named because it's a hybrid originally developed in Trinidad with the aim of imbuing high-quality *criollo* with *forastero*'s fortitude. The *criollo* genotype is the only Cacao cluster to be found growing wild in Central American forests, while all ten genotypes—including *criollo*—occur in South America, suggesting that human agents took an ancestor of *criollo* Cacao from the Amazon basin in South America and transported it north of the Andes.[6]

At some point in the distant past, people discovered the use of Cacao seeds to make drinks through fermenting, toasting, grinding, and mixing them with water. Cultivated *criollo* is the result of their efforts: they bred Cacao to keep fruiting for more years, and to produce seeds that taste better and give a good buzz, but

Immature Criollo pod 3, and flowers,
@ San Antonio Suchitequepez

made it more dependent, lower-yielding and less adaptable in the process.[7] *Criollo* seeds are ivory or pale purple in cross-section because they contain fewer anthocyanins[i] and more caffeine than other varieties, making them less sour and more stimulating.[8]

Traces of theobromine, a chemical compound found in Cacao, and starches from Cacao seeds have been detected on ancient ceramics. The earliest definitively Cacao-tainted relics in Central America are fragments of broken pottery dated to 1900 BCE,[9] discovered in the Cacao-cultivating district known as the Soconusco, west of Lake Atitlan on the Pacific coast of Guatemala and Mexico. Similar evidence has been discovered in present-day Ecuador (in South America) from 3400 BCE,[10] so Cacao's agricultural symbiosis with man was verifiably underway in South America over five thousand years ago. Even earlier origins for its domestication and export up into Mesoamerica seem likely: Mesoamerican ballcourts, usually coeval with later, Cacao-centric agrarian cultures such as the Maya and Mexica, have been dated from 5000 BCE,[11] significantly antedating earlier estimates of the transition from hunter-gatherer to settled societies in Mesoamerica.

Criollo may be the most delicate and disease-sensitive variety of Cacao tree, but it's also the longest lived, with a lifespan of a hundred to one hundred and fifty years.[12] Wild *criollo* Cacao grows on riverbanks, and in some *dzonots* or *cenotes* of the Central American Yucatan peninsula, sink-holes that penetrate deep below the surface and expose the water table, providing perfect warm, damp, and sheltered microclimates which benefit Cacao's insect pollinators. *Criollo* Cacao was principally cultivated in the Soconusco and present-day Tabasco (see Figure 1) on the Mexican east coast, because the climate in these regions perfectly suits the plant.[13]

---

[i] Although lower in anthocyanins, *criollo* Cacao is higher in caffeic acid derivatives than other varieties, so there is little net loss in "antioxidant" capacity (Elwers *et al.*, 2009); I speculate that there may even be some overall gain in medicinal utility.

Figure 1. Principal Cacao cultivation zones in Mesoamerica

Food crops were traditionally grown alongside Cacao in Mesoamerica in "companion planting" arrangements. One such ancient system came to be known in colonial times as the *milpa* ("maize field") method, where several different types of plants for food, medicine or domestic use were grown in the same plot, each supporting the other—for example, beans would use maize stalks to climb up, while bacteria in their roots fertilised the soil by fixing atmospheric nitrogen, essential for plant growth.[14] The Yucatan Maya mostly grew Cacao in the traditional way—in what the Spanish later named *cacaotales*, 'Cacao orchards': biodiverse, multicropping systems, like large *milpas* established on high ground for better drainage. More than six hundred and twenty-five Cacao trees could be grown on a hectare of land, alongside other crops and useful plants.[15]

Vases from the Early Classic Maya Period in Mesoamerica (250–600 CE) depict feasts with large cylindrical pots of frothy *kakawa*, a Mayan name for Cacao-based beverages; other representations show liquid being poured from one of these vessels into another from a height for the purpose of creating a head of foam. By the Late Classic period (600–800 C.E.) the Maya began using spouted vessels for serving *kakawa*, and specific pottery vessels for different types of *kakawa* have been identified, often labelled with hieroglyphic inscriptions describing their contents, such as 'tree-fresh *kakawa*' (a beverage made from Cacao fruit pulp, perhaps) or 'honey *kakawa*' (an early form of sweet chocolate drink). Some inscriptions contain instructive descriptions or phrases, such as '*takan kel*'—to roast Cacao well, in order that beverages made from it would produce a lot of '*yom kakawa*', or foam. Some of the containers were of an ingenious lock-top design, having a twist-on lid with a handle to facilitate the transportation and storage of pre-prepared *kakawa*.

Traditional Mesoamerican Cacao-based beverages are made from fermented seeds. After harvesting the pods, the seeds are scooped out and piled up, and microbes in the traces of fruit pulp sticking to the beans multiply and cause chemical changes. This process takes anything from a few hours to several days, depending on variables such as local custom, ambient temperatures, and the desired flavour of the beans (longer fermenting results in a sourer, more wine-like complexity of flavour). The seeds are then sun-dried, toasted, and shelled; they are traditionally ground on a sun- or coal-warmed three-footed portable stone table called a *metate*, sometimes with other ingredients. They liquefy due to their high fat content, and this so-called cacao liquor may be allowed to cool and set into tablets or discs for storage and later use; if fresh ingredients are incorporated, the resulting liquid or dough can be mixed with hot or tepid water immediately, and the finished drinks are frothed by pouring from a height or beating with a whisk to raise a head of foam.

The Maya consumed both hot and cold *kakawa* potions, but, like cocoa-drinkers today, they mostly ingested Cacao in hot drinks. The much later Mexica (the Aztecs), who achieved cultural predominance in Mesoamerica from 1300–1521 CE, drank their *cacahuatl* at room temperature—which in the valley of Mexico is tepid, not cold. In other respects, Mexica and Maya Cacao drinks were basically similar—mostly bitter or savoury, spiced, and foamy concoctions with many variations.

The highest class of Cacao drinks prepared by the Mexica, known as *tlaquetzalli* [tul·ack·et·zall·ee] *cacahuatl*—referencing the iridescent turquoise plumage of the *quetzal* bird, signifying 'precious thing'[16]—all had an enormous head of foam. In fact, the drinks may have been at least 50% froth, to be eaten with a spoon: botanical foaming agents were used, many of which are still employed in the multifarious Central American beverages known as *atoles*, corn-based gruels which serve as staple foods in many regions. Such foaming agents may have included:

- Fresh aromatic Frangipani flowers (*Plumeria rubra*), known to the Mexica as *cacacolxochitl* [ka·ka·loh·zo·tchul] and as *cacalosúchil* today;
- Fresh, lightly toasted stems of a young Sarsaparilla species, probably *Smilax aristolochiifolia*, known as *cozolmecatl* [koz·oll·meh·cat·ul] to the Mexica, and referred to as *cocolmeca* or *azquiote* in contemporary Mexico;
- The fruit peel, the fresh root inner bark, or latex of a vine related to the Milkweeds of southern USA and known locally in contemporary Tabasco by various names, including *chupipe* and *cahuayote* (*Gonolobus niger*), and the fresh inner stem bark of another *Gonolobus* species on the east coast, possibly *G. barbatus*, known locally as *n'ched*;
- Anaerobically fermented seeds of *pataxtle* or *pataxte* (*Theobroma bicolor*), a close relative of Cacao, buried in water-filled pits for six months then shade-dried, colloquially referred to as *cacao blanco* or *cacao címarron* in Oaxaca today.

There were (and are) many more plant-based foaming agents, often region- or culture-specific. It seems likely that the most esteemed pre-conquest Cacao drinks were filtered, then foamed by agitation, turning the entire volume into a light, aromatic froth; or the foam was collected separately and added to the top of the beverage. Other libations may have been made

Cymbopetalum penduliflorum UNAM herbarium specimen 2018

richer and denser with the addition of fatty additives, such as cooked *gaucoyules* or Corozo palm tree seeds (*Attalea cohune*) and toasted, crushed *mamey* seeds, the almond-scented pits of *Pouteria sapota* fruits. When ground into a paste with Cacao and aromatic additives, the lipids in these seeds precipitate in aerated, whipped-cream-like clumps atop the liquid as water is slowly poured in from above, as some pre-conquest illustrations illustrate, and as is currently performed in modern-day Oaxaca in the manufacture of the regional *atole* known as *tejate*.

The favoured chocolate spice of the Mexica was Ear Flower or *xochinacaztli* [shoh·chee·na·kaz·tlee] (*Cymbopetalum penduliflorum*), known today as *orejuela*. The plant was also known as *hueinacaztli* [way·na·kaz·tlee] ('great ear'), and *teonacaztli* [tay·oh·na·kaz·tlee] ('divine ear'), because the leathery dried flower petals with a complex, resinous bouquet look— with a little imagination—a bit like ears.[17] Two other plants of almost equal pre-Colombian prestige are now global superstars. Vanilla was *tlilxochitl* [clil·zoh·tchul] or 'black flower'[18] to the Mexica, because after harvesting the ripe seed pods of the climbing, vine-like orchid, they are scalded, carefully fermented for several months so they blacken, and dried to develop their unique aroma.[19] Chilli fruits (*Capsicum* spp.) also originate from Central America, despite their ubiquity in Asian cuisines—a testament to their pseudo-masochistic culinary appeal and medicinal utility; they were a common ingredient in Mesoamerican medicines, foods, and drinks.[20]

The Funeral Tree (*Quararibea funebris*) was second only to Ear flower in its traditional significance as a Cacao-drink spice. It's trumpet-shaped yellow flowers are called *Rosita de Cacao* ('little rose of Cacao') in contemporary Mexico, and the Mexica knew them as *cacahuaxochitl* [ka·ka·wah·zoh·tchul], or 'flower of Cacao'. They have a floral scent somewhat reminiscent of Vanilla which is remarkably durable—hundred-year-old dried specimens in Mexico City University's herbarium retain their powerful perfume undiminished. Allspice berries, so named because they reminded Europeans of a blend of Old World aromatics, such as cinnamon, nutmeg and cloves, were known as *xocoxochitl* [shok·oh·zoh·tchul] to the Mexica, and also have a long history of use in Cacao-based beverages.

Another ingredient, this time used primarily for its colour rather than its flavour, was *achiote*, a Hispanicised version of *achiotl*, a Nahuatl name referring both to the seeds of the native plant known as annatto (*Bixa orellana*), and to the brick-red cakes of food dye made from them. This little tree has blousy pink flowers, which morph into burgundy seed pods like frondy Martian testicles. The carmine seeds are soaked in hot water to extract the colour, then the blood-red liquid is evaporated over heat until it thickens; maize flour or fat and salt and other spices are added to solidify and preserve the concentrate, which may be dried into round red patties for storage and use in cooking, known today as *pasta de achiote*.

Holy leaf (*Hoja Santa*, or *Piper auritum*), called *mecaxochitl* [meh·ka·zoh·tchul] by the Mexica, is botanically related to Asian black pepper (*Piper nigrum*) and the Polynesian anaesthetic, kava-kava (*Piper methysticum*). The Mexica and the Maya preferred the leaf for cooking and medicine, but used the dried flower spikes in *cacahuatl*. Popcorn flowers (*Bourreria huanita*), known to the Mexica as *izquixochitl*[21] [is·key·zoh·tchul] have a Jasmine-like odour when fresh, a wonderful flavour combination with Cacao, and *yolloxochitl* [yoh·low·zoh·tchul] (*Magnolia mexicana*) flowers are bright yellow when dried, with a strong melon-like scent; known as *yolosúchil* today, they are alleged to have been the favourite *cacahuatl* spice of emperor Motecuhzoma III, aka Montezuma, the ill-fated last ruler of the Aztec empire.[22] All three plants have fallen out of use as Cacao spices in contemporary Central America.

Following the Spanish takeover in the early 16th century, after an initial aversion to the bitter, heathen, strangely non-alcoholic Cacao-based beverages, the colonists gradually acquired a taste for *cacahuatl*. They began to consume Cacao in hot drinks, in the manner of the Yucatec Maya. They added sugar to sweeten it; they substituted most native spices (with the notable exception of Vanilla) with Cinnamon from Sri Lanka, Black Pepper from the Indies, and Aniseed from Europe; and they elaborated the *molinillo*, a beautiful type of wooden hand whisk for frothing the drinks, still in use in Mexico today. I suggest elaborated, not invented, because there is second-hand evidence that the Mexica used some kind of utensil for frothing *cacahuatl*: an inventory of ingredients and equipment for a Mexica banquet includes "chocolate beaters ... two or four thousand of them".[23] The Spanish also changed the name of the drink: *cacahuatl* became *chocolate*. The consumption of this sweet exotic nostrum in European courts popularised the drink, and consequently, over the next two hundred years or so, the Old World appropriated Cacao.

Cacao gained territory globally after the Spanish conquest, but lost ground in its homeland. At the time of the Spanish conquest, a Cacao orchard such as Izalco in El Salvador had 33,570 trees each producing 400,000 beans per tree, annually.[24] Only a few years later, land mismanagement, indentured slavery of the native population, over-taxation and corruption had reduced Izalco's output to almost nothing. The colonists almost exclusively used slave labour to cultivate Cacao, which increased the flow of Cacao to Europe and through the upper echelons of colonial society at the cost of human suffering and diminishing returns from traditional Mesoamerican sources, obliging Europeans to seek new locations to cultivate Cacao in the West Indies and South America.[25] Perhaps not coincidentally, flowers from a wild-type Central American *criollo* Cacao variety today—genetically closer to ancestral native cultivars—have a chemically complex aroma which draws more midges than the less sophisticated scent of hybridised trees in contemporary Cacao farms[26] other words, there is a marked decline of fertility in cultivated *criollo*, which may be due in part to post-conquest farming techniques. Similarly, in modern times, the discovery of Cacao's mycorrhizal symbiosis suggests that the use of anti-fungal sprays may have unforeseen detrimental effects further down the line.

By the 1800s, consumption of chocolate drinks in Europe had declined in popularity. Coffee was cheaper, and unlike Cacao, which must be ground to a liquid consistency before being made into a drink, coffee and tea can simply be infused in hot water. In 1828, the Dutchman Coenraad van Houten invented the process that came to be called 'Dutching'—treating unroasted, shelled and fragmented cocoa beans or nibs with alkali to make the final product more miscible with water, then roasting and pressing them to extrude the fat, or cocoa butter, and grinding the resulting de-fatted roasted seed cake to powder. The result of this process is cocoa powder—darker, dryer, and lighter than Cacao bean mass (so-called cocoa liquor), much easier to mix with water, but also markedly less flavoursome.

In 1847, the Joseph Fry invented the world's first chocolate bar, by combining cocoa powder, cocoa butter, and sugar; and when Henri Nestlé produced evaporated (powdered) milk in 1867, it was only a matter of time before someone took the next logical step, and that someone was the Swiss manufacturer Daniel Peter, who in 1879 became the first person to oversee the production of edible milk chocolate, the delectable combination of cocoa liquor (Cacao bean paste), milk powder, sugar, and extra cocoa butter. The final step in the transformation of Cacao from bitter beverage to sweet confection—and another step down in potency from the handmade plant-based drinks of ages past—was the invention of the conching process in the same year by Rudolphe Lindt. Conching is the term applied to the repeated beating of cocoa liquor by mechanized rollers over many hours or days to reduce any particles in the liquid to the smallest possible size and make it as smooth as possible,

Tikal, 2011

Young Cacao tree, Tapachula, Mexico 2018

so that the finished chocolate has a completely grit-free and orally pleasing texture.

So, Cacao, the sacred seed which was the basis of various elaborate beverages for the New World elite, became a commonplace pleasure of the Old World masses: a sweet, secular snack. As noted by a 17th century colonial physician and Cacao enthusiast, Dr William Hughes, the inclusion of extra ingredients

> "makes the Chocolate never the better; and without such addition, it is excellent good, and very agreeable, strengthening Nature exceedingly [...] And truly, what we now use in England, is but a compound of Spices, Milk, Eggs, Sugar &c., and perhaps there is in it a fourth or fifth part of the chiefest ingredient, the Cacao [...] So it is no wonder if this Drink be not found of that virtue and efficacy as hath been noised abroad, or as many expect: But doubtless if Physitians did but narrowly pry into the secrets of the nature of it, they would quickly finde (the right use thereof being made) that it can scarcely be too much commended" (Hughes, 1672).

So, let us pry.

CHAPTER TWO

Pharmacologies

"**pharmacology** (n.) "the sum of scientific knowledge concerning drugs," 1721, formed in Modern Latin (1680s) from *pharmaco-* (see pharmacy) [...]

pharmacy (n.) [...] from Greek *pharmakeia* "a healing or harmful medicine, a healing or poisonous herb" [...] from *pharmakon* "a drug, a poison, philter, charm, spell, enchantment" [...]

-logy: word-forming element meaning "a speaking, discourse, treatise, doctrine, theory, science," from Greek *-logia*."

(Online Etymology Dictionary, 2023)

i. Ancestral *Kakawa*

Mesoamerican medicine recognised four elements (water, earth, fire, and wind) comprising the world, each relating to a season and a compass direction—almost identical to the coetaneous European humoral scheme. Unlike Old World theories deriving from ancient Hellenic and Medieval medicine with their four qualities of heat, cold, damp, and dryness, the Mesoamerican model was binary, proposing only heat and cold as

Molinillos

primary states.[ii] The Mexica believed that the part of the soul that gave reason and individuality, *teyolía*, dwelt in the human heart, and that there were three primary causes of disease: natural, magical, and divine.[27] All illness was seen as disharmony, reflecting a person's relationship with society, the gods, the ancestors, or the environment, so divining the root cause of imbalances—often supernatural, such as a failure to pay proper respects to ancestors, or behavioural—was most important.[28]

The Tzotzil Maya conceived of an invisible force or immaterial magical substance they called *itz*, the "blessed substance of the sky", which flowed out of gateways to the otherworld opened by the *itzam* or shaman. *Itz* was metaphorically described as, and coalesced through ritual action into resins, blood, sap, nectar, honeycomb, milk, sweat, tears, semen—and Cacao. These are all natural substances which were seen as repositories of life force, or *ch'ulel*. *Itz* was transformative, an invisible substance that caused change and generated new possibilities; the stuff of magical action, growth and transformation in the world. *Ch'ulel* was a more fundamental and ethereal force, less dense than *itz*, often visualised as vapour: in space, as nebulae amongst the stars; in the sky, as clouds bringing rain; above the altar, in the smoke of burning incense and offerings; and in living blood, where it carried the spiritual essence of the ancestors. In a sense, *itz* may be the product of ritually precipitated *ch'ulel*.[29] For the later Mexica, the breath of life, or *yoliliztli*—the equivalent of Mayan *ch'ulel*, perhaps—was concentrated in the heart. One of the Mexica poetic metaphors for Cacao was *yollol eztli*, or 'heart, blood', with *yol* being a common root in the word *teyolía*. So, for the Mexica *yoliliztli*, the animating

---

[ii] Although, it should be noted that in the Galenic-humoral schema, heat and cold were described as active qualities, whereas moisture and dryness were passive. Assuming that the records of indigenous medical theory aren't tainted by post-colonial historical revisionism, the philosophical overlap between the historical Eurasian and Mesoamerican models is remarkable, considering they are presumed to have evolved independently on separate continents.

force of the world—the heart of reality—was linked to Cacao, or *yollol eztli*, the blood of the heart.[30]

Mexica poets repeatedly speak of Cacao as beverages as exhilarating. The "intoxicating" properties of "flowery" (foaming, precious) Cacao were associated with singing, dancing, splendour, and celebration of life, despite a continual sorrowful awareness of inevitable death.[31] Cacao-based drinks were served during the making of treaties and contracts; this use of Cacao to seal a deal is depicted in the 13th century post-classic era Mayan Madrid Codex, where the elderly rain god Chac and Ix Chel, the young moon goddess of fertility and healing, exchange gifts of Cacao seeds over a foaming goblet of *kakawa*.[32] This use could infer some pharmacological properties which may be useful during negotiations, or perhaps imply nothing more than Cacao's valued status as a drink, analogous to the serving of champagne at social functions.

Mesoamericans regarded Cacao as a cooling and nourishing tonic, to satisfy hunger and thirst, and to protect the complexion from sun damage.[33] The Mexica incorporated Cacao in treatments for physical ailments such as fevers, skin eruptions, diarrhoea and bloody dysentery, indigestion and flatulence, lung problems, exhaustion, impotence, to delay hair growth, to clean the teeth, and as a prophylactic against snakebite. Cacao was a key part of magico-medical remedies for what we would now consider to be mental health issues, such as agitation, vision-quest hangovers, and insanity, and was used to assist or prohibit sleep, or to increase courage and reduce fear.[34] The Mexica are also known to have drunk *cacahuatl* at the end of their feasts, when psychoactive *Psilocybe* mushrooms were often consumed;[35] similarly, there are accounts of indigenous men in Nicaragua shortly after the Spanish conquest using Cacao to recover from a shamanic ritual in which some unidentified plants were smoked in a pipe, accompanied by dancing and drumming. The men "fell senseless on the ground or ran amok weeping … and had to be carried off to bed by their wives or

Foaming kakawa 2023

friends".[36] In the aftermath of this psychic assault, they were left suffering from "stupidity of mind" for a week or so, during which they were dosed with Cacao-based beverages.

Although post-colonial influences are inevitably present, contemporary folk uses of Cacao are rooted in ancient traditions. The Nahua of Mexico drink chocolate as an aid to weight gain in wasting diseases, and Cacao is perceived as beneficial in hot weather and when feeling overheated—hence its use as a tonic (often combined with other, more specific plants) in febrile states. In post-conquest Nicaragua, it was said that drinking chocolate in the morning would provide some immunity from the ill-effects of snakebite in the afternoon.[37] In present-day Oaxaca in Mexico, drinking chocolate is thought to help relieve symptoms of bronchitis, and even afford some protection from wasp, bee, and scorpion stings.[38] Other twentieth-century uses of Cacao seeds and chocolate from the Dominican Republic include anti-anaemic properties, to improve kidney function, and to "ease the brain when overexerted".[39] These applications all accord with pre-Hispanic characterisations of the plant.

Cacao also has an important and often unremarked role in folk obstetrics today. In the Mexican state of Morelos, drinking chocolate is used on its own as a parturient—to help a mother give birth—which makes sense, as a stimulant with some anti-anxiety and pain-relieving properties could be quite useful.[40] Cocoa butter, the fat extracted from Cacao seeds, is used by skilled midwives in Guatemala: when an unborn child is in breech (with its feet rather than its head pointing towards the exit), a traditional midwife can diagnose the problem by manual palpation. She then greases the woman's belly with cocoa butter and uses deep abdominal manipulation to turn the baby around.[41] Cacao is also used to promote lactation in breastfeeding mothers, either taken as drinking chocolate or consumed in the form of an *atole* made with maize and Cacao for at least two months to increase the milk supply.[42] An additional benefit of chocolate here, whether for new mums or not, is for wasting or malnutrition, and drinking it is seen to protect a woman when "the body is wasting, [it] has nothing, no food".[43]

When Cacao was first brought to Europe, the dominant medical model in the Old World was humoral medicine, also known as Galenism—named after its godfather, the Roman physician and author Claudius Galen, who wrote several books laying down the theory and ground rules for the practice of medicine. Galenic medicine recognised four elements, these being earth, air, fire and water, each represented in the human body by a fluid, or humour. These fluids could be described as a group of functions: blood contained the largest quantity of the air element, and was known as the sanguine humour; it was warm and moist (like air), and nourished the body. Melancholy or black bile had a greater concentration of earth, and was therefore cold and dry, and helped retain necessary substances. Hot, dry choler or yellow bile incorporated a lot of fire, which helped with digestion, appetite, and motivation; while cool and moist phlegm contained a lot of water, and its function was to lubricate the body and remove waste products.

Un-processed raw Cacao seeds were thought to be predominantly cold and dry (melancholic), although they had other qualities, too; the dominant cold, dry, earthy part was counterbalanced by an oily, warm and moist part, and a very hot and dry component.[44] Processing Cacao seeds by toasting then mixing them with spices, water, sugar, or other ingredients changed the overall temperament of Cacao-based beverages, so separating the humoral nature of Cacao itself from the interference created by other ingredients in drinking chocolate would be a challenge, even if the entire system weren't a non-standardised, empirical and subjective art. Whatever its nature, a moderate intake of Cacao in the form of drinking chocolate was thought to support the vital spirits—life energy, or vitality, residing in the heart—and, unusually for something made from a cooling plant, it was reputed to expel excess melancholy, a surfeit of which could cause depression, stiffness, anorexia, or constipation.[45]

ii. Contemporary Cacao

The melancholy humour was associated with sourness, sanguine with sweet or fatty tastes, and choleric with bitterness; so, the cold and dry part of Cacao seeds may be analogous to what we now recognise as astringent polyphenols. The fat in Cacao seeds may be the warm and moist part, and the stimulating alkaloids, including caffeine, may be said to comprise the bitter-tasting hot and dry elements. Over half the weight of fermented, dried and toasted Cacao seeds is composed of fats and related substances, with approximately 15–25% carbohydrates, including fibre such as cellulose and pectin, and naturally occurring sugars. These fillers with little direct pharmacological activity may still affect the properties of the seed once ingested, by modifying the absorption of other constituents, or providing food for gut microbiota. Cacao contains significant levels of copper and magnesium; a single 40g serving of Cacao contains more than the entire daily nutritional requirement of copper for an adult.[iii] The more pharmacologically "active" material—the alkaloids, polyphenols, amino acid-based compounds, and organic acids—comprise somewhere between one third and one fifth of the seeds' dry weight altogether (see Figure 2).

The fats or lipids in Cacao collectively have the useful property of being liquid at body temperature but solid below 25°C, enabling chocolatiers to produce chocolate that melts in your mouth (or your hand, or your pocket, etc.) Just over half of that is stearic acid, a type of saturated fat which lowers 'bad' LDL cholesterol. The fatty portion of the seeds also contains a small but appreciable quantity (about 0.2%) of phytosterols, hormone-like compounds which reduce intestinal cholesterol absorption; they may also modulate immune responses,

---

[iii] Based on an RDI of 0.9mg, Cacao contains 24ppm copper, equating to 0.96mg per 40g. (ppm values from Dr Duke's Phytochemical & Ethnobotanical Database online, https://phytochem.nal.usda.gov/phytochem/search).

Winnowing toasted cacao, Rex'hua, Guatemala 2011

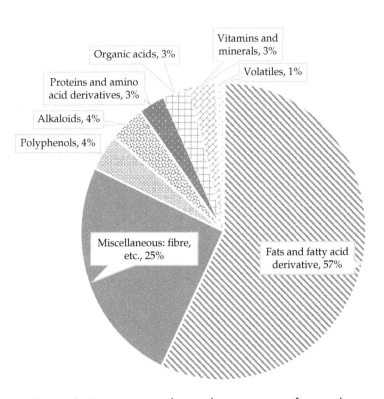

Figure 2. Approximate chemical composition of toasted
Theobroma cacao seeds, as % dry weight

help to prevent several types of cancers, and reduce the immediate stress effects of hard exercise on the body.[46]

Much has been made of Cacao's antioxidant properties, resident in the polyphenols, a group of compounds which collectively comprise about 4% of the toasted beans' weight; chocolate contains the highest weight of polyphenols in any food.[47] The greater proportion of Cacao's polyphenols is made up of relatively inert lignin, a form of indigestible fibre which helps to feed good bacteria in the gut, and has been shown to contribute to gut flora variations in mice.[48] The remaining compounds, including procyanidins and flavanols, neutralise free radicals (unpaired electrons produced during ordinary cellular metabolism, like soot from a fire), reducing tissue damage or wear and tear; they also promote release of nitric oxide from the linings of blood vessels, causing them to dilate, thereby improving blood flow.[49]

Cacao, once eaten, significantly reduces lipid oxidation in the blood, which is generally a good thing: this indicates that Cacao most likely counteracts *atherosclerosis*, or 'furring up' of arteries and blood vessels, and may therefore reduce the risk of developing heart disease. A 2011 meta-analysis of Cacao's usefulness for preventing cardiovascular disease (CVD) concluded that people who eat the most chocolate were 37% less likely to have a heart attack and 29% less likely to have a stroke than if they ate little or no chocolate: smokers, the obese, people who didn't exercise, and people who ate only fried food all had similar reductions in relative risk if they ate more chocolate.[50] A 2022 meta-analysis similarly found robust reductions in absolute risk of death in men consuming more chocolate over a thirty-year period, with a 16% reduction in risk of death from heart disease, 13% from CVD, and 12% from cancer at higher levels of dietary intake.[51]

Cacao also appears to be beneficial for prevention of diabetes mellitus and metabolic syndrome. The mixture of flavanols in Cacao rapidly lower the levels of triglycerides in the blood and reduce fat production (lipogenesis) in mice at high levels of intake.[52] Human studies have substantiated these results: in a majority of trials, high-polyphenol dark chocolate improved pancreatic function and insulin sensitivity, modestly reduced blood pressure and cholesterol, and improved arterial blood flow acutely or over two weeks, in doses equivalent to 74–100g dark chocolate daily, with stronger effects noted for sugar-free chocolate.[53] More recent lab and animal research suggests that Cacao may help protect both the liver and pancreas, reducing the risk of developing diabetes and fatty liver disease.[54]

Pertinent to Cacao's traditional reputation as a roborant or fortifying substance, a single 50g serving of dark chocolate reduced stress responses in a small group of men aged twenty to fifty years old, with blunted serum cortisol and adrenaline spikes in response to social stress a couple of hours after eating it.[55] Likewise, thirty days' intake of 40g dark chocolate per day reduced urinary levels of cortisol and breakdown products of adrenaline, and this effect was greatest in people with higher levels of anxiety.[56] High stress levels, associated with elevated blood cortisol, are associated with a shorter lifespan. With excessive stress we literally age faster, so perhaps unsurprisingly, a retrospective review of a group of 470 chocophilic men aged sixty-five to eighty-four from the Netherlands between 1985 and 1990 found a 47% reduction in relative risk of death from all causes put together during the study period (compared to non-chocolate consumers).[57] A more recent US population study involving more than nine thousand participants aged fifty-five to seventy-four found a 13–16% reduction in relative risk of death from all causes for those eating more than half of one 28.35g serving of any type of chocolate—milk or dark—per week![58] And eating chocolate (of various kinds) only one to three times a month has been correlated with a whole extra year of life in population studies.[59]

This protective effect of Cacao goes further: Cacao polyphenols as a whole, when consumed regularly, prevent or delay the development of several types of cancer in

non-human animal experiments at levels of intake compa-
rable to those which are achievable with traditional Cacao
beverages in people.[60] Cacao polyphenols lower levels of
oxidative stress marker compounds in the bloodstream,[61]
assuage inflammation in the lining of blood vessels and else-
where,[62] and neutralise or blunt the effects of several carcino-
gens.[63] The transformation of healthy cells into cancer cells
is increased by all of these factors (stress, oxidation, poor
blood circulation, inflammation, or cancer-causing chemicals). Added to the diet of rats, Cacao gradually increases
the general immune response to viruses and cancers,[64] and if
this effect is replicated in humans, then it may provide more
evidence that Cacao is chemopreventive, a substance which
reduces the risk of developing cancer.

One of Cacao's less dramatic traditional indications is to protect the skin against sun damage. Two hours after consuming
Cacao, blood flow to the skin increases by 70%.[65] A twelve-week
double-blind study testing the effects of high-flavanol cocoa
in a group of twenty-four women found that those women
ingesting the equivalent of 30–40g whole Cacao per day had a
significant reduction in signs of sun (UV light) damage such as
reddening of the skin, with measurable improvements in skin
texture and hydration, although, sadly, no changes in wrinkles
were noted.[66] Another thirty-person double-blind trial con-
firmed that high flavanol cocoa increased the skin's resistance
to UV damage over a twelve week period.[67]

The primary alkaloids in Cacao (alkaloids being compounds containing nitrogen and are alkaline) are the three
so-called methylxanthine alkaloids—theobromine, caffeine,
and theophylline, plus a sprinkling of other trace compounds.
Of these, *caffeine* is the most stimulating—acting by block-
ing receptors for adenosine, a signal-damping brain chemical
or neurotransmitter, causing stimulatory and mood-elevat-
ing effects. Cacao contains around 0.2% caffeine,[68] but a 40g
dose of Cacao seed which may be consumed in a traditional

Cacao-based beverage contains 84mg of caffeine, greater than the amount in a cup of tea.

At 1.2% of the seed's weight, *theobromine* is the most plentiful alkaloid in Cacao., and it's also produced in the body as a breakdown product of caffeine. Theobromine is a stimulant, but people's sensitivity to it varies much more than with caffeine; it may even decrease caffeine's stimulating properties, although it makes no difference to caffeine's positive effect on mood.[69] Theobromine's actions on the heart (increasing the rate and force of the pulse) and the kidneys (increasing urination) are stronger than its effects on the brain, and it dose-dependently increases heart rate.[70] It also has the interesting property of strengthening tooth enamel.[71] Theobromine weakens some cancer cells—most strikingly, a highly aggressive form of brain cancer called glioblastoma—and helps to protect blood vessels.[72] In this regard, its actions complement those of the polyphenols.

Habitual dark chocolate consumption has been linked to improved cognitive function relative to abstainers, bestowing better abstract reasoning, working memory, and visual-spatial memory on chocolate eaters.[73] As with the findings in the cardiovascular studies, these chocolate-related benefits were independent of other factors, such as exercise, fruit and vegetable consumption, obesity, and general health. Isolated theobromine improves memory task performance in rodents, and increases cyclic adenosine monophosphate (cAMP) signalling in brain cells, potentially enhancing the expression of brain-derived neurotrophic factor (BDNF), a compound that promotes learning and memory. A recent, small randomised human study showed that thirty days' consumption of 24g of 70% Cacao dark chocolate, but not the same quantity of white chocolate (made with only cocoa butter plus milk and sugar, so lacking Cacao's alkaloids and polyphenols) resulted in elevated blood levels of theobromine and nerve growth factor during the trial period. The dark chocolate group had improved performance in cognitive tests which was still measurable three weeks later, after blood levels of theobromine had returned to zero.[74]

Cacao also has a long history of use for pain control. At minimum, chocolate's reputation as a palliative food or treat could reinforce a placebo response, which would diminish pain, but the polyphenols and methylxanthines in Cacao also have measurable pain-reducing effects. Caffeine alleviates several types of pain, and the flavanols suppress inflammatory responses by reducing production of inflammatory compounds and improving circulation,[75] such that consuming a flavanol-rich dark chocolate drink after exercise diminishes next-day stiffness and discomfort.[76] Cacao flavanols specifically reduce both inflammation and pain perception in *trigeminal neuralgia*, a type of facial pain which can be a result of nerve damage during dental work.[77]

Cacao's traditional reputation as beneficial for some lung ailments is corroborated—incredibly—in the finding that children born to mothers who consume chocolate during pregnancy have a lower incidence of asthma.[78] A laboratory study using whole Cacao extract showed that it reduced the production of a substance called neopterin in white blood cells, indicating anti-inflammatory effects and the potential for Cacao to affect or modulate the immune system, perhaps damping down some types of allergic responses.[79] Similarly, the polyphenol compound clovamide dilates the airways in vitro with the same potency as the anti-asthmatic drug salbutamol (often marketed as Ventolin™). Clovamide was detected in the urine two hours after humans consumed Cacao beverages, indicating that it found its way into the general circulation, and might be expected to affect the lungs;[80] it reduces coughing caused by inhaling chilli fumes, for example.[81] The alkaloid theophylline in Cacao even more strongly inhibits airway spasm and inflammation, as well as impairing histamine release;[82] likewise, theobromine suppresses the cough reflex and slightly dilates the airways, effectively reducing persistent coughing. In children, the airway-dilating effects of both theo-bromine and theophylline last for two to six hours after inges-tion,[83] so these effects are likely to be present in whole Cacao.

Cacao also has a traditional reputation as a prophylac tic against the harmful effects of poisonous bites and stings. In addition to its value as a pain reducer or anodyne, we know that Cacao may diminish clotting, dilate blood vessels and reduce inflammation in blood vessel linings, whereas the venom of many poisonous snakes increases blood clot-ting and inflammation in blood vessels, causing circulatory blockages, pain, and tissue death. Some of the procyanidins in Cacao inhibit an enzyme called hyaluronidase, which breaks down hyaluronic acid, a component of healthy tissue.[84] This hyaluronidase inhibitory activity of Cacao's procyanidins may be another reason for its traditional use as a general or supportive tonic in many illnesses including cancer and lung ailments, and possibly its use as a prophylactic antivenom against some types of snakebite—because many snake ven-oms, too, contain hyaluronidases, so impeding this enzyme's activity with plant extracts significantly reduces morbidity (ability to cause suffering and harm) in experimental animals.[85] But Cacao hasn't yet been tested for this purpose (any volun-teers?), so this is currently an intriguing speculation.

There is a sprinkling of trace alkaloids in Cacao, each com-prising a fraction of a percentage of the seed. These include salsolinol, which may interact with the brain's endorphins and modify the release of a hormone called prolactin, which brings on lactation in breastfeeding mothers, perhaps speaking to Cacao's traditional use as a galactagogue.[86] Also present are the tetrahydro-beta-carboline alkaloids (or THβCs), and trigo-nelline. Like salsolinol, THβCs are products of fermentation which occur in many other foods, and can be manufactured in the human brain. THβCs and salsolinol have been found to strengthen opiate and alcohol dependency in primates and rats in lab experiments[87] perhaps by activating dopamine and opiate signals in the brain,[88] and several THβCs inhibit the breakdown of the "happy" brain chemical serotonin, although the precise activities of the THβCs in Cacao are unknown. Trigonelline is present in Cacao at very low levels, and is produced from ingested niacin (vitamin B3, also found in Cacao seeds), and may lower cholesterol and blood sugar.[89]

Proteins and related compounds comprise around 11% of the fermented, toasted and dried seeds. Proteins and their constituent amino acids—the bricks from which proteins are made—mainly have nutritional value. Amino acids themselves are made up of amines, many of which are used as chemical messengers in the brain. Those present in Cacao include dopamine, serotonin, tyramine, phenethylamine (PEA), and histamine. Tyramine and PEA have both been described as neuromodulators—they modify the effects of other brain chemicals on mood, perception, and behaviour. Their effect when eaten all mixed together in food is thought to be negligible, although their relative quantities and the presence of other compounds can affect how they are absorbed and distributed in the body. Tyramine, dopamine, and PEA are broken down in the body by enzymes—naturally occurring catalysts—called monoamine oxidases (MAO). Several compounds in Cacao inhibit this type of enzyme, which could allow these amines to build up; and the trigonelline in Cacao can ship PEA directly into the central nervous system (the brain and spinal cord), where even low tissue concentrations of PEA increase the motivational effects of dopamine and other excitatory (stimulating) chemicals.[90]

Cacao also contains a small amount of gamma-amino butyric acid (GABA), which has anti-anxiety and sedative properties in adults.[91] Toasted Cacao seeds also contain diketopiperazines (DKPs), compounds with a bitter taste which contribute to Cacao's overall flavour, but very little is known of their physiological effects, except that they can cross the blood-brain barrier.[92] Compounds called amino alcohols are also present, and these include anandamide, which reduces pain, anxiety, and fear, as well as promoting weight gain and controlling milk production during breastfeeding.[93] Like Cacao's amines, anandamide occurs naturally in the human brain and binds to cannabinoid receptors—it's the natural substance which is mimicked by compounds from some *Cannabis* species, and is often referred to as an endocannabinoid (the prefix *endo-* comes from a Greek root word meaning within). It's active at tiny doses,[94] and the other amino alcohols in Cacao, the *n-acylethanolamines*, inhibit its breakdown at very low levels.[95] Aromatic alkaloids called *pyrazines*, produced during roasting, especially in *criollo*,[96] may enhance the effects of Cacao's other constituents on circulation and cognition.[97]

I refer to this cocktail of amines, endocannabinoids, trace alkaloids and DKPs as Cacao's "fairy dust" compounds. The quantity of each compound is exceptionally small, but tiny doses don't always mean tiny results—if they make it into the brain, or combine their effects with other chemicals, or build up over time, then effects may be disproportionate. For example, the following constituents of Cacao are known to have monoamine oxidase inhibitory (MAOI) properties:

- Inhibiting MAO-A (causing elevation or preservation of serotonin, noradrenaline, dopamine, melatonin, adrenaline, and tyramine): proanthocyanidins/anthocyanins, salsolinol, resveratrol, THβCs.
- Inhibiting MAO-B (causing elevation or preservation of phenethylamine, serotonin, tyramine, dopamine): proanthocyanidins/anthocyanins, the flavan-3-ols catechin and epicatechin.

Similarly, many compounds in Cacao activate opioid (endorphin) receptors, such as the flavanols epicatechin and catechin, which bind to delta- (δ-) opioid receptors on heart cells.[98] Most of the polyphenols in Cacao don't bind directly to opiate receptors in this way, but they do stimulate the release of the gaseous compound nitric oxide (NO), which in addition to dilating blood vessels also affects opioid and dopamine pathways in the central nervous system. Several of the polyphenols in Cacao (such as flavan-3-ols and procyanidins) are known to have this effect.[99]

As we have seen, Cacao was consumed alongside *Psilocybe* mushrooms by the Mexica, and traditionally used to help recovery from what we would now call "altered states". In recent times, chocolate has been employed to accelerate the onset of MDMA effects in group therapy,[100] resorted to by opiate addicts as a means of prolonging or augmenting the high,[101] and even consumed by magic mushroom users with the expectation that it will amplify the effects of the mushrooms.[102] Apart from such historical and anecdotal evidence, there is some preliminary scientific support for Cacao affecting brain chemistry in a way that implies synergy or modulation of the effects of other psychoactive substances. For example, daily consumption of 50g dark chocolate for one month increases serotonin metabolism in humans,[iv,103] and surveys found that people with depression were more likely to crave chocolate when depressed, and to eat more of it.[104] Perhaps a key finding is that both a single (one-off) dose of Cacao containing 450mg flavanols, equivalent to a serving of 40g whole Cacao beans in a traditional drink, and a third of this dose taken every day for five days significantly increased blood flow to the brain as measured by an MRI scan in human volunteers,[105] suggesting that Cacao may be an effective "drug delivery" tool to the central nervous system for other substances which affect brain chemistry.

A handful of small human clinical trials may also be relevant here. In one trial, "mindful" consumption of chocolate improved mood much more than "non-mindful" consumption of chocolate or of dry crackers, whether they were eaten mindfully

---

[iv] As measured by increased plasma levels of 5-hydroxyindoleacetic acid, or 5-HIAA, a serotonin metabolite in the human trial participants.

or not;[106] and in a second trial, where the effect of eating chocolate on mood was compared to drinking plain water, chocolate
improved the mood of participants but only if they actively
focused on feeling better when eating it.[107] In a more unconven-
tional experiment, half-ounce (14g) servings of 68% cocoa solids dark chocolate were "blessed" by either a shaman, Buddhist
monks or a group of meditators and eaten twice daily on three
days in one week by volunteers, who were instructed to savour
each piece and primed to expect "optimal health and function-
ing" and "an increased sense of energy, vigor and well-being"
from the chocolate. In this double-blind placebo-controlled trial,
the blessed chocolate produced a significant improvement in
mood relative to un-blessed chocolate, and this effect was stron-
gest in those who didn't regularly eat much chocolate.[108] Taken
together, these preliminary experiments suggest that chocolate
can elevate mood more than other foods, but that this potential
may require activation through *intention*.

* * *

According to the research of Dr Helen Fisher, formerly of
Rutgers University, New Jersey, it seems that love and sexual
attraction can be placed into three broad neurochemical cat-
egories, based on lust (sex drive), infatuation (romantic attraction), and bonding (attachment, or comfortable familiarity).
Infatuation is a more unstable state characterised by elevated
levels of the excitatory neurotransmitters dopamine, noradrenaline, and PEA, and low levels of serotonin—the latter two of
which may be modulated by Cacao. Bonding is driven by two
neurohormones—hormones which also affect nerve cells in
the brain—known as vasopressin and oxytocin. Animal behavioural research shows that infants which are given consistent
maternal care, resulting in regular boosts of oxytocin release,
have lower cortisol (stress hormone) levels and less intense
stress responses, stronger immune systems, and themselves

become more nurturing as adults—in other words, they may be better at bonding, and less prone to infatuation. Curiously, enhanced activity of nitric oxide in the brain (as experienced following ingestion of Cacao) is thought to be crucial in "love", perhaps especially in bonding, because it "helps to keep or facilitate a state of calmness and contentment … resembling morphine signalling" (Esch & Stefano, 2005).[109]

Nitric oxide acts on the main bonding hormone oxytocin and on endorphins (also involved in infatuation) to consolidate or diminish pleasure responses, depending on the circumstances. The brain's natural endocannabinoid, anandamide—also present in, and perhaps boosted by Cacao—triggers release of nitric oxide in specific neurons in the posterior pituitary gland, the control organ for sex hormones, to switch off vasopressin and oxytocin release, while it also stimulates oxytocin release through a separate pathway.[110] Activation of serotonin receptors (5HT1A) in the brain elevates oxytocin levels, too; Cacao not only contains serotonin-sparing (MAOI) compounds, but we've seen evidence from animal studies that larger quantities of Cacao may raise serotonin levels in the brain, and that regular chocolate intake measurably alters serotonin metabolism in humans.

A small feeding trial in rats showed that one-off ingestion of Cacao—approximately equivalent to a modest serving of 13g whole Cacao for an adult male of average weight in the UK[v]—had no effect on instinctual fear reactions, but significantly reduced conditioned fear responses, even more than low doses of a standard tranquilliser drug (an intravenous benzodiazepine). Conditioned fear arises from the pairing of an initially neutral stimulus to an aversive stimulus, which causes sentient organisms (a human or non-human animal) to anticipate an

---

[v] Calculated using 83.9kg as UK average adult weight, and dividing the dose of Cacao used in the trial rats (100mg/100g) by 6.2 to attain a Human Equivalent Dose, as recommended in 2005 FDA guidelines for clinical trial dose estimation.

Grinding Cacao on the metate, 2019

aversive event (punishment) through association. This is a behavioural response acquired from associative learning from often harsh or unpleasant experiences, that become linked to triggers, such as a backfiring car causing a war veteran to duck under cover. This differs from instinctual fear responses to such things as snakes or spiders, which are hard-wired, primordial reactions designed to keep us safe. Conversely, when fed to the rats daily over two weeks, the same dosage of Cacao slightly increased conditioned fear reactions, accompanied by a mild rise in brain serotonin levels.[111]

These findings all raise the possibility that Cacao—and its sweet product, chocolate—may have the capacity to affect individual perceptions and responses, and perhaps by extension to influence wider social behaviours and predilections. Corroborating this theory, a study of 305 new mums who experienced antenatal stress reported more negative temperaments (fear responses) in their children at six months old; but mothers who ate chocolate weekly or daily during pregnancy rated their children's temperaments more positively at six months postpartum than mums who ate no chocolate, and the more chocolate the mothers ate during pregnancy, the more positively they rated their children's temperaments.[112] A survey of 1367 elderly Finnish men of similar income level and social status, averaging seventy-six years old, discovered that chocolate eaters had "better health, optimism and … psychological well-being".[113] Similarly, in a small survey of sixty-five young US university students, chocolate preference was found to be greater in the subset who believed their destiny to be predominantly under their own control than their less chocolate-inclined, more fatalistic fellows.[114] None of these psychosocial studies tells us whether Cacao consumption is a consequent to, causal of, or just coincident with their findings; but the correlations are intriguing, and suggestive.

* * *

The Mexica's correlation of *yolliliztli* (the life force in the heart) with Cacao is eerily echoed by contemporary discoveries about the beneficial effects of Cacao's polyphenols on the circulatory system and metabolism, and the epidemiological correlations between Cacao consumption and increased lifespan (relative to chocophobic Cacao-avoiders). It's possible that ingesting real, traditionally prepared Cacao-based beverages—or even good quality dark chocolate with a high cocoa content—may help to prevent or assuage many of the most significant health challenges of the affluent world. Cardiovascular disease, diabetes, some types of cancer, and many age-related problems such as cognitive decline, chronic pain, or respiratory conditions—even, to some extent, the aging process itself—can potentially be ameliorated by the judicious use of Cacao, albeit preferably in combination with significant lifestyle and dietary adjustments.

The Maya concept of Cacao as a repository of *itz*, or potential for change, may speak to state-dependent effects of Cacao on the neurophysiology of stress and pleasure. Specific interactions between Cacao's many constituents and the brain chemistry associated with infatuation and bonding may even infer a broader capacity for subtle modulation of social interactions in chocolate-consuming populations. In pharmacological lingo, Cacao is certainly a stimulant owing to its caffeine and theobromine content, but hasn't yet been shown to be an antidepressant. I suggest that the truth is more nuanced: Cacao may be a *hedonic modulator*, amplifying or adjusting pleasure responses, and an *anti-phobic* agent, attenuating conditioned fear responses. Likewise, Cacao appears to modify the effects of other psychoactive plants or drugs, and enhances the influence of intention on positive mood states.

How can we know what ultimate effects Cacao, or any other significant medicinal or food plant, has on the individuals and societies which partake of it? Speculation about how Cacao affects human beings and humankind is more likely to be fruitful and accurate if the subject is examined in three ways: Historically, to provide a framework, or body of information (Chapter 1); scientifically, to give it a rational, mental dimension (Chapter 2); and metaphysically, because the myths and rituals of Cacao speak to its character or spirit as perceived by society, to which we now turn.

CHAPTER THREE

Ceremonies

> "**ceremony (n.)** late 14c., *cerymonye*, "a religious obser-
> vance, a solemn rite," from [...] Latin *caerimonia* "holi-
> ness, sacredness; awe; reverent rite, sacred ceremony"
> [...] Also from late 14c. as "a conventional usage of
> politeness, formality." The disparaging sense of "mere
> formality" is by 1550s."
>
> (The Free Dictionary, 2023)

Cacao was deeply rooted in Mesoamerican mythology. The Axis Mundi is a myth found in many cultures: a giant tree connecting this world with the underworld below and the celestial world above. The Maya referred to it as the "croco-dile tree", because when the *wakah-chan* (the Milky Way, an astral representation of the Axis Mundi for the Maya) lay parallel to the horizon at night it became the "cosmic crocodile" with two giant starry "jaws". Cacao trees are sometimes depicted as crocodile-trees on Maya pottery.[115] Cacao is also illustrated emerging from the jaws of a serpent, representing a conduit between this world and the otherworlds, which can be accessed in visionary states—an appropriate association for a tree that grows on the edges of cenotes, those natural portals to the underworld.

The theology of the Mesoamerican post-mortem experience—a shuffling of souls from the human world of the sun, to the underworld, or to the celestial world, and back again, is incompletely understood. It seems that some aspects of the soul could be reincarnated in different forms, appearing in our sub-solar realm as plants or animals or humans. Especially important and venerated ancestors were often shown returning as Cacao trees, or their spirits as being somehow accessible through them, and Cacao trees were depicted growing right out of ancestors' graves.[116] Cacao was also strongly associated with femaleness and fertility in Mesoamerica,[117] and this tradition continues with its folk medical use by Guatemalan *comadronas* (midwives). The pendulous pods growing directly from the Cacao tree's trunk resemble testicles or breasts, and Classic Maya artworks and effigies depict women sprouting Cacao fruits, or covered in Cacao seeds, while a late Classic Maya male figurine has a split Cacao pod on a stalk emerging from the top of his loincloth, like a penis (though a museum catalogue describes it, tactfully, as emerging from his "navel").[118]

In one scene on a late Classic Mayan vase, a person is shown emerging from the base of a Cacao tree, rising up towards a man with a resplendent headdress. Perhaps the man with the headgear is a king or a shaman-priest invoking a royal ancestor, as he stands welcoming the revenant with open arms; another man with chocolate-stained lips sits on the ground, looking up at the seated ruler-god K'awil, who gestures towards the Cacao tree, while a servant grinds Cacao seeds on a *metate* at his feet. Here, the ritual use of Cacao correlates with resurrection or ancestor-worship, and royal power.[119] Similarly, relief carvings of Cacao trees flank the entrance to the Temple of the Owls—messenger birds of the underworld—on two standing stones at the late-Classic Maya site known as Chichen Itza. The graven trees are dotted with representations of jade disks, denoting preciousness, while fully three-dimensional carved figures emerge from the base of the trees, arms crossed around their chests, with facial motifs indicating that they are breathing and speaking or singing.

Young 'Cacao god' in the Popol Vuh museum, Guatemala, 2011

Serpentine, umbilical "tails" connect them to the trunk of the Cacao trees.[120] Given their location, it seems likely that these are representations of ancestor-spirits being brought to life through Cacao.

The K'iche Maya myth of the Hero Twins combines many of these themes. Adult twin brothers play a noisy ball game, which disturbs the gods of the underworld beneath the earth. The twins are invited to play a ball game against the gods of death, and are tricked into losing, forfeiting their lives, where-upon their bodies are buried at a crossroads in the world of the dead, and one of the men's heads is hung up in the branches of a tree by the side of the road as a gruesome trophy. The tree begins to bear fruit, which attract the attention of Blood Moon, the daughter of the king of the underworld; but when she approaches the tree to pick one of the fruits, the severed head spits in her palm and she becomes pregnant, because myth is no respecter of biology. Terrified of how her tyrannical father will react to her improbable fecundity, she flees to the human world and seeks refuge with the twins' mother, where she eventually gives birth to twin boys.

This second generation of twins grow up, and the cycle repeats—they play a noisy ballgame, disturb the gods of death, and are invited to the underworld. Before they leave, they plant corn in their grandmother's garden, and tell her to watch it: so long as the corn lives, they live. They play a game against the gods of the underworld, and this time around, with a little help, they win; but instead of honouring their promise to reward them, the gods of death plan a murderous betrayal. Fore-warned, the twins leave instructions that their remains are to be ground to powder and dumped in the river in the land of the dead, then throw themselves into a fire before the gods of death have the chance to kill them. Their pulverized ashes are scat-tered into the river, while above ground the corn they planted dies, and their grandmother mourns. But a few days later, the corn re-sprouts, and two fish appear in the underworld river.

The fishes emerge from the water, transforming into human miracle workers, who tour the underworld killing and resurrecting the souls of the dead. Intrigued, the gods of death invite them to their court to perform this miracle, not realising that the pair are the reanimated second generation of twins. The twins decapitate the gods of death and refuse to reanimate them, before visiting their father and uncle's remains at the crossroads, and finally exiting the underworld as the sun and moon.[121]

This entire tale can be read as a complex allegory for processing Cacao seeds to make beverages. The Mayan glyph for *kakawa* was two fish, because the word *ka* meant "fish"[122]—so *ka-ka-wa* is a pun. The second generation of twins' entrance to the underworld and the trials they undergo are equivalent to fermentation; they jump into the fire, and are roasted; their bones are ground, like Cacao on the *metate*; then their ashes are scattered into the water of an underworld river so that they can reincarnate as two fishes, just as Cacao is mixed with water to produce *kakawa*.[123] The Hero Twins are the future sun and moon, linked to maize and Cacao—of the elder twins, one dies and one is decapitated (perhaps representing the harvesting of a maize cob or a Cacao pod, or an eclipse) then reborn as the fruit of a tree in the underworld; of the younger generation, again one is decapitated, both die as a maize plant dies, and both are reborn in the underworld as two fish. The sun-maize-Cacao association in Mesoamerican thought has been extended to a sun-maize-Cacao-Christ overlap in contemporary Mayan mythology: modern day K'iche even say a blessing over Cacao seeds in Christian ceremonies, dedicating them to "the resuscitation of our Lord"[124]—in this case, the persecuted, sacrificed, and resurrected Christ.

In the Hero Twins myth, the elder twin's pod-head impregnates Blood Moon (a red moon being an eclipse phenomenon), who later bears the Hero Twins, living embodiments of maize and Cacao, and the sun and moon. A scene from the post-Classic

Maya Madrid Codex shows a young moon goddess, perhaps Ix Chel, the Mayan goddess of healing and fertility, exchanging Cacao beans and sharing a pot of foaming *kakawa* with the elderly rain god Chac.[125] *Kakawa* was drunk when making alliances and important agreements, and this scene symbolises fertility in the wider sense of agricultural fecundity, dependent on the rains and—according to Mayan beliefs—the moon's cycles. Cacao has ancient use in rain and fertility rituals, both as an offering and a sacrament, and is strongly associated with water,[126] the element which is paradoxically evocative of both the underworld and fecundity, because moisture is a precondition for growth and decay. So, in various Mesoamerican myths and traditions, Cacao may represent both male and female fertility and the cyclical rhythms of life, death, decay, and rebirth.

Quetzalcoatl, the feathered serpent, was the great benefactor of humankind for the Aztecs. In Nahuatl, the language of the Mexica, *quetz* means spirit or wind; the *quetzal* is a native bird with blue-green iridescent plumage which was prized more highly than gold in Mesoamerica, while *coatl* is the word for snake.[127] So Quetzalcoatl's name could be translated as "luxuriously feathered wind-spirit-bird-snake". Quetzalcoatl descended into the underworld to retrieve the bones of the people from the previous world of the Fourth Sun which he had ruled, and had ultimately been destroyed by flood. Quetzalcoatl then gives the bones to the old goddess Cihuacoatl, who grinds them "like cornmeal" on a *metate*, adding blood from Quetzalcoatl's penis to create a kind of living clay from which the first humans of the Fifth Sun were made.[128] This occurred in the human origin-place called Tamoanchan, which is represented in codices as a flowering tree cut in half and gouting blood.[129] While this myth doesn't explicitly reference Cacao, Cacao and maize were paired essentials in the iconography and diet of pre-Colombian Mesoamerica, often ground together on a metate as a prerequisite for making Cacao-based *atollis*; and in the Mexica Fejérváry-Mayer Codex, Cacao is the tree associated with the south cardinal direction, the "wind leaving place" or gateway to the underworld,[130] and we know that both Cacao and human blood were potential repositories of magical *itz* for the ancient Maya.

Cacao was historically associated with wealth, status, and luxury, as well as their covert corollaries of slavery and subjection in both pre-Colombian Mesoamerica and European colonies from the early modern period onwards. The world's biggest consumers of chocolate are not the growers of Cacao, but perhaps it was ever thus: in parts of Mesoamerica where Cacao could not grow it was ingested by nobles, royals, and on special occasions, and great quantities were traded or paid in tribute to cities like Teotihuacan and Tenochtitlan, far to the north of its zones of cultivation. Now, Cacao is mostly cultivated in Africa, Asia, and South America, but predominantly consumed in Europe. When contemplated from the torch-lit twilight of magical thinking, perhaps it may also be said that following the European colonization of Mesoamerica, Europe was in turn colonised by Cacao; by 2021, annual chocolate consumption in Switzerland was 11.6kg (25lb 9oz) per person, with the US in second place at 9kg (19lb 13oz), followed by Germany, then France, and the UK in fourth place.[131] The story of the plant is curiously bound up with themes of privilege, luxury, hierarchy, and serfdom, even to the present day.

In one Mexica origin story for Cacao, a princess looks after her warrior husband's estate and hoarded treasure while he is away on a military campaign. She is kidnapped and tortured by her husband's envious enemies to extract the location of the treasure, but reveals nothing. When she finally succumbs to death, the benefactor-god Quetzalcoatl causes the Cacao tree to emerge from her blood: the fruit "hides the real treasure of the seeds", which are pink-tinted, strong, and bitter—representing blood, virtue, and "the suffering of love".[132] Implicit in this myth are connections between Cacao and the life force present in the blood, and perhaps most

Sra. Delifina Valverde makes atole for a traditional Cacao
beverage, San Mateo del Mar, Oaxaca, 2018

explicitly, Cacao is linked to hidden wealth as envy-inciting earthly riches, or, more positively, to the transformative spiritual treasures of courage and loyalty. So Cacao has symbolic links to blood, rulership, women, wealth, the underworld, fertility, water, the heart, the head, and the "rebirth" of gods and ancestors; it's mythically paired with maize, and the cycles of the sun and moon; it is sometimes represented as the world tree, connecting the human world to the otherworlds above and below.[133] To drink *kakawa* is to replenish *ch'ulel*, perhaps augmenting fertility in the wider sense of recharging one's ability to acquire *itz* and generate magical change in the world.

Cacao ceremonies in their current form are a modern-day invention pioneered by Keith Wilson, the charismatic self-styled "chocolate shaman" who resides in the town of San Marcos by lake Atitlan in Guatemala. Keith researched Cacao and used visionary techniques to "commune with the Cacao spirit" to devise his ceremonial procedure, which has since been imitated, adapted and embellished around the world, often without attribution.[134] Keith's process involves administering moderately large doses of Cacao—43–45g as a starting dose—to all the adults present, then talking all participants into a meditative state where they focus on their bodies, and zone in on any area of pain, discomfort, or blockage, aiming to release it. The ceremonies are a "safe space" where participants are encouraged to cry, laugh, move around, or do whatever to release any blockages they encounter under Keith's watchful eye. The effects are obvious—thirty minutes after everyone drinks their Cacao, the amplifying effect of the guided meditation becomes apparent, with increased excitement, euphoria, mood changes and in some cases nausea and physical discomfort; many people became visibly emotional, some laughing, some crying, others sitting as if transfixed. An optional top-up of 25–30g Cacao is ingested after an hour or so.

By contrast, while Cacao had ritual importance in Mesoamerica, few details of how and why it was incorporated into historical rituals are known. Some Mayan cultures used *kakawa* as a baptismal liquid, mixing water taken from hollows in trees or rocks in the forest—natural fonts, or miniature portals to the watery underworld—with Cacao and certain flowers to make a special anointing chocolate,[135] which was painted onto the forehead, face, and spaces between the fingers and toes with a bone.[136] In this ritual context, chocolate symbolises new life, and its use in baptism symbolically connects the infant to his or her ancestors in the underworld, and the cycle of life. In contemporary Belize, food and drinking chocolate are left in caves as offerings to underworld deities;[137] and in Guatemala, cave offerings to the earth gods and angels consist of animal blood, Cacao, and fermented beverages,[138] demonstrating old associations between the earth, food, fermentation, the underworld, and Cacao.

Page one of the pre-conquest *Codex Borgia*, produced by a vassal nation of the Aztec empire under Mexica rule, illustrates the Mesoamerican equivalent of the big bang: an explosion of stars and serpents pouring out of the top of a large turquoise bowl, presided over by a black, skeletonised being with clawed hands and feet. In Mesoamerica, snakes represented rebirth and transformation,[139] rulership and authority, lightning—including the "lightning in the blood" experienced by diviners and shamans, sudden revelations of truth revealed through physical sensation[140]—and desirable but dangerous knowledge. Snakes were also described as the "companions of rainbows"[141], and vision snakes are depicted in Mayan art, where the snake's maw was the *yol*, an open portal between realms created during visionary rituals;[142] snakes have been used as descriptors for the "living umbilical cord" or conscious pathways between realms.[143]

The star-studded snakes in the *Codex Borgia* illustration have been interpreted as outflows of power evoked by the Nahuatl couplet *yohualli ehectal*, night-wind.[144] This ophidian eruption resembles a dark head of foam effervescing out of a skull-faced bowl, from which one might consume a traditional Cacao beverage: are the origins of the universe depicted here as a cauldron of *cacahuatl*? It's notable that when Cacao beverages are made without added maize and properly frothed, a film of natural oils in the foam catches the light, so that every bubble contains a swirling miniature rainbow, like iridescent quetzal plumage. In Mesoamerican terms, the foam on a cup of revitalizing *kakawa* or *cacahuatl* is inhabited by thousands of tiny living rainbow-serpents, reminiscent of the creation-drawing in the *Codex Borgia*, with snakes streaming out of the froth in the cosmic skull-cup.

Because Cacao itself was frequently used as an oblation, many offering pots were decorated with Cacao pods.[145] The Maya sometimes used non-perishable analogues of blood, such as ground hematite or cinnabar,[146] and prepared *kakawa*—perhaps coloured blood-red with annatto—may have constituted an offering of this type. Contemporary Yucatec Maya still place prepared Cacao beverages on an altar so that the spirits can drink their "essence"[147]—their *ch'ulel*, perhaps, or its equivalent. Cacao's association with rebirth and reanimation links it to life in the broader sense of summoning spirits or deities into objects or statues in rituals, and ensouling them. The Lacandon Maya traditionally placed five Cacao beans, symbolising the heart, lungs, liver, stomach, and diaphragm into their "god pots", which were subsequently ensouled in a ritual to house their gods;[148] the Cacao beans are the vital organs of the idols, without which the gods cannot inhabit them.

Cacao's association with the heart is also metaphorically tied to its property of instilling courage. This can be seen in its historical use as a libation for sacrificial victims, warriors, and merchants—those who have faced, or are about to face death—and in its mythical associations with valorous acts, such as the Mexica princess's bravery, Quetzalcoatl's blood sacrifice to re-create humanity, and the courage of the Hero Twins. The heart is the first of the five vital organs symbolised by Cacao beans in the "god pots" of the Lacandon Maya.

To remove a heart is to extract life-force, so perhaps Cacao's synonymity with the heart and the blood identify it as a roborant at the level of the soul.

Cacao reduces levels of adrenaline and the stress steroid cortisol, and may influence many of the neurohormones involved in the subtypes of love known as infatuation and bonding.[vi] So it's perhaps unsurprising that a survey of chocolate consumers found greater sensitivity to negative emotion among people who ate the most chocolate.[149] Taken together with traditional portrayals of Cacao as an agent of ancestral rebirth or transformation through adversity, it seems reasonable to speculate that those drawn to Cacao may encounter, embody, or expose inherited patterns of trauma; perhaps chocoholics are unconsciously attempting to self-medicate such tendencies. When used intentionally, Cacao may have the potential to be a medicine for helping to steer this process in more constructive and less self-sabotaging directions. As the Cuban proverb says, "toma chocolate, paga lo que debes"—"drink your chocolate, pay what you owe".[150]

* * *

Cacao's pharmacology suggests that it may be a hedonic modulator with anti-phobic properties and at least partly intention-dependent effects. *Entheogen* is a term typically applied to powerful psychoactive plants with shamanic or religious uses, and simply means "realising the divine within".[151] If basic meditation or visualisation are likened to something between a trip in a hot air balloon and a spacewalk in earth's orbit, then entheogens are more like rockets blasting those who partake of them into outer (or inner) space, regions of consciousness which may otherwise be accessed only by years

---

[vi] See last chapter, or Chapter 5 of *The Secret Life of Chocolate* for a more thorough exploration of this possibility.

of rigorous meditative or ritual discipline, or perhaps more precipitously in dream or delirium. In this analogy, perhaps Cacao could be compared to a useful amenity kit containing a spacesuit, a jet pack, and a lunchbox: something which makes expeditions into inner worlds through ritual, meditation and/or the use of entheogens more accessible, and a little bit safer.

It may be more accurate to call Cacao a proto-entheogen, or an entheogen-enabler. The historical records of Cacao's use alongside psychedelic mushrooms provides some corroboration for this hypothesis, as does the archaeological association of Cacao with portals to the otherworlds, and ritual invocation of ancestors or transformative magical forces. With these speculations in mind, here are some ideas for incorporating Cacao into ceremonies:

1. Keith Wilson's ceremonies utilise Cacao's putative antiphobic properties to facilitate what may be described as *somatic releasing*: feeling into the body and helping to dissolve blocks. Cacao may therefore be experimented with in settings or rituals designed to help overcome fear-based psycho-physical obstacles, preferably with a tried and tested process, ceremonial structure or therapeutic method. Cacao may prove to be useful in therapeutic or ceremonial processes addressing PTSD, as chocolate was once taken alongside MDMA—since found to be highly effective for this disorder[152]—in therapy sessions to amplify the effects of the drug. If Cacao is used for this purpose, or to address other serious psychiatric issues, this should be done under the supervision of a qualified psychotherapist.

2. David Pearl's Street Wisdom process involves some simple mental exercises not dissimilar to mindfulness, and boils down to going on a walk with an open mind to help find answers to life questions—a sort of guided DIY divinatory experience.[153] Given Cacao's traditional use as a ritual facilitator and psychic lubricant, and its possible nootropic or

cognition-enhancing actions, it seems highly suited for use in similar immersive inspiration-unlocking processes.

3. Cacao's adrenaline, and cortisol, lowering capacity, its property of increasing blood flow to the brain, and its traditional ceremonial use as a conduit for or repository of *ch'ulel* or *yolliliztli* all suggest potential value in meditation and visualisation. Anecdotally, Cacao does enhance focus and relaxation (i.e. concentration) in meditation.

4. People have reported that consuming Cacao in the evening gives rise to vivid, brightly coloured, and intense dreams.[154] These experiences suggest that Cacao may be useful in lucid dreaming, a key technique in many magical traditions. This is a very short list of therapeutic and esoteric possibilities to explore in ceremony with Cacao; there are many more that may be derived by further contemplation and experiment.

A word of caution is warranted: Cacao is, in one sense, a tool. Like other powerful plant medicines, Cacao can be conceptualised by the metaphysically-inclined as a magical substance, or an entity in its own right; but just as when working with a human therapist or a medical device, no external entity or contrivance will do all the work for you—ultimately, there are no shortcuts. In other words, becoming a chocoholic (or a Cacaoholic) won't solve any problems! Given a choice between carving rock with your bare hands or using a chisel, tools are preferable; similarly, fighting one's demons is easier with a supportive ringside coach equipped with towel, water, bandages, and so on. Cacao isn't a panacea, nor is it a solution; but it may be a kind of psychic solvent, helping to temper and mitigate problems as part of a larger strategy of resolution.

The pharmacology and mythology of Cacao suggest that it may be a valuable facilitator of de-traumatisation and transformation. At the very least, Cacao is a mostly beneficial, low-toxicity substance with potentially vast public health benefits—and it's certainly a pleasant medicine to try.

Abuelita selling Cacao, Rosita, Mamey y Achiote,
Mercado 20 Noviembre, Oaxaca city, 2008

Formulary

Theobroma cacao L.

<u>Plant part used</u>: Fermented and toasted seeds.

<u>Pharmacological actions</u>:
- *Clinically demonstrated*: adrenaline-lowering; anti-cariogenic; anti-ischaemic, anti-thrombotic; anti-tussive; anti-ulcerogenic; astringent (mild); cortisol-reducing; nutritive; insulinotropic; stimulant; vascular endothelial anti-inflammatory; vasodilatory; weak hypotensive.

- *Inferred from pre-clinical and/ or epidemiological research*: anti-phobic—attenuates conditioned fear (acute/ short-term effect); anti-senescent; bronchodilator (mild); chemopreventive; coronary vasodilator; enteric microbial trophorestorative; galactagogue (weak); hedonic modulator—amplifies or adjusts pleasure responses; mild analgesic; nephroprotective; nootropic; tolerogenic.

Therapeutic indications (abbreviated—complete list in *The Secret Life of Chocolate*, Appendix A):

- Short-term or acute:
  - a) *Clinically demonstrated*—cough, fatigue, high-intensity exercise (to aid recovery), low mood (with intention), potentiating Paracetamol's analgesic effects, sleep deprivation, trigeminal neuralgia.
  - b) *Hypothetical (inferred from pre-clinical and/or epidemiological research)*—angina, cramp, DVT, IgA-mediated food sensitivities, intermittent claudication, portal hypertension in liver disease, PTSD[vii] sickle cell crisis, social anxiety.

- Longer-term preventive or supportive:
  - a) *Clinically demonstrated*—atherosclerosis, chronic fatigue syndrome, CVD and CHD prophylaxis, gestational hypertension, hepatic cirrhosis, hypertension, insulin resistance and metabolic syndrome, UV skin damage.
  - b) *Hypothetical (inferred from pre-clinical and/or epidemiological research + historical usage)*—ALS (amyotrophic lateral sclerosis), angina, asthma, cancer prevention (especially prostate and colon cancer), chilblains, COPD, Cushing's disease, dementia, diabetic vasculopathies, dysbiosis, erectile dysfunction, glioblastoma—especially with Doxorubicin (synergist), peptic ulceration, post-ischaemic recovery (e.g. post-stroke, heart attack or DVT), prostate enlargement associated with benign prostatic hyperplasia (BPH), renal failure, Reynaud's phenomenon and Buerger's disease, snakebite—prophylaxis (may attenuate), sickle cell disease or thalassaemia, smokers—skin damage, varicose ulcers, vascular dementia, weak tooth enamel.

[vii] As an adjunct to psychotherapy.

Contra-indications (avoid, or use with caution):

- High risk (avoid): Allergy to chocolate or Cacao.
- Medium risk (avoid, or monitor closely): acne, especially in adolescent males; fibrocystic breast disease; gastro-oesophageal reflux disease (GORD or GERD), or heartburn; post-menopausal women with low bone density; Wilson's disease.
- Uncertain or hypothetical risk (monitor closely): Addison's disease; kidney stones; MAOI antidepressants—with higher acute/ceremonial doses of Cacao (e.g. moclobemide, phenelzine, isocarboxazid, tranylcypromine); male infertility—low sperm count; migraine headache; Parkinson's disease (may benefit); tyramine sensitivity; urticaria.

Dosage:
- Acute dosing or ceremonies, adults—whole Cacao only:
  - Single doses of 40–50g, supplemented with up to two top-up doses of 20g at 90-minute intervals.
  - Dark chocolate, cocoa powder and Cacao powder are not recommended for acute dosing owing to their lower pharmaceutical quality (pharmacological complexity is significantly impacted by processing).

- Longer-term, adults, supportive use—adjust in each case according to tolerance and effects:
  - Whole Cacao: 10g daily, or 20–40g three times per week.
  - As dark chocolate (minimum 85% cocoa solids), approximately 10–15g daily or 20–50g thrice weekly.
  - As un-Dutched cocoa powder, 10–20g daily, or minimum three times per week.

- Children: Cacao isn't suitable for the very young, and larger ceremonial doses are not recommended for children under 16.

Guidelines (adjust as appropriate according to growth pattern, bodyweight, sensitivity etc.):

- ○ ages 5–8, one third to one half of the adult dosage;
- ○ ages 9–12, from two thirds to three quarters of the adult dose;
- ○ 13–16 years of age, up to the full adult dosage.

Preparations

In terms of Cacao's medicinal potency and pharmacological complexity, the hierarchy of available Cacao-based products—with some variation, depending on sourcing and processing differences—is likely to be:

1. **Home-made cocoa liquor** (home-toasted, shelled, and ground cocoa beans). Cocoa liquor made from beans which have been lightly toasted then shelled and ground "by hand" (or at home) are likely to contain a good spectrum of Cacao's pharmacologically active constituents.
2. **Ceremonial Cacao** blocks: shelled, roasted, ground and cooled cocoa liquor. Usually toasted at lower temperatures, and may be made from higher quality beans. Provenance, processing and quality varies widely.
3. **Cocoa nibs** (roasted and shelled but unground pieces of Cacao seed) are liable to have been roasted at higher temperatures, degrading (oxidising) more polyphenols.
4. In addition to this slightly higher roasting temperature, industrially produced blocks of set **cocoa mass** or couverture are likely have been *conched*, i.e. ground by granite rollers for many hours which mellows flavour but reduces many volatiles and slightly denatures the Cacao.
5. **Dark chocolate** is also made of conched Cacao, plus added fat and sugar, further "diluting" the pharmacology of the bean. 85% cocoa solids is the minimum suggested strength if the chocolate is intended to have beneficial effects, as "cocoa solids" includes added cocoa butter, and the remainder is sugar. Artisanal bean-to-bar chocolate is

often much higher quality than big-name factory-made chocolate.

6. **Raw Cacao powder**, although theoretically higher quality than dark chocolate or nibs, has been de-fatted, making the polyphenols it contains more susceptible to oxidation in storage once the bag is opened. Raw Cacao also lacks many of the under-researched compounds produced during toasting, such as the diketopiperazines.

7. **Cocoa powder** ordinarily has been roasted and de-fatted, so is even less pharmacologically complex. "Ceremonial grade cocoa powder" is a tautology.

8. **Dutched cocoa powder** has been treated with alkali to saponify the remaining fats and make it more miscible with water, which ever further oxidises polyphenols. Both ordinary and Dutched cocoa powders are also more likely to be higher in cadmium and lead than other forms of Cacao.[155]

For making Cacao beverages at home, *criollo* Cacao from Central American Cacao-growing countries (Mexico, Guatemala, Honduras, El Salvador, Costa Rica) is recommended for optimal flavour (less bitter/sour), better subjective effects (more caffeine, more caffeic acid derivatives, and more procyanidins than anthocyanins appear to result in more marked mood alterations), and maximal historical authenticity—even though, strictly speaking, Cacao ceremonies in their current form aren't historically authentic. *Fermentado* (three or more days of fermentation) or *lavado* (up to twenty-four hours of fermentation) Cacao should be selected, and truly raw beans—if they exist— may be best avoided, as they taste bad and lack many of the "fairy dust" constituents produced during fermentation.

Serving suggestions for beverages:

1) Mexica style *cacahuatl*: use real Cacao (homemade cocoa liquor, nibs or couverture.[viii] Simple version: blocks of ceremonial

---

[viii] See *The Secret Life of Chocolate*, Chapter 8, for bean-to-beverage recipes.

Cacao or couverture may be chopped or grated, and weighed out into a jug. Spices, powdered herbs and (optional) sweetening may be added, then boiling water. 60–120ml boiling water per 20g Cacao—less water for a thicker, more intense beverage, more for a thinner drink, or if more powders (spices or medicinals) are added. Beat very well with a whisk, molinillo or coffee frother to dissolve the Cacao, disperse any powders, and raise a good head of foam; pour and serve.

2) *Atole* con Cacao—very versatile, easy to prepare.[ix] Simple version: sift a tablespoon or two of organic *masa harina* flour into a pan, add water and whisk while bringing to a boil to make a plain gruel with a soup-like consistency; use this *atole* to make a beverage with grated Cacao, spices and sweetening, as above. Makes a thicker, more savoury, non-foaming Cacao-based drink.

3) Smoothies—use nibs, grated Cacao and/or powders in smoothies, e.g. with cooked bananas (raw banana contains high levels of polyphenol oxidase, which will lower polyphenol levels), dates, oats, nuts, beans, and/or spices. Avoid using animal milk or soya milk if possible as the proteins in these milks may reduce bioavailability of some of the polyphenols in Cacao.

For qualified Medical Herbalists:
NOTE: *Incorrect dose or use of medicinal plants can be dangerous, so combining medicinal plants with Cacao (other than the most innocuous food-grade plants) should be done with the oversight or input of a qualified medical herbalist.*
Herbal combinations:

1. Finely powdered herbs and spices can be ground with Cacao beans if making beverages from scratch (Mesoamerican style). Alternatively, they can be added to drinks or smoothies just before whisking or blending, as Cacao beverages are viscous and will hold powders in suspension.

[ix] See *The Secret Life of Chocolate*, Chapter 8, for recipes.

2. Hot herbal infusions can be used to make Cacao-based beverages.
3. Liquid herbal preparations such as aromatic waters, tinctures, or even essential oils (if safe for consumption, and at low concentrations) can be added to Cacao preparations.

Limitations include the need to avoid adding very bitter herbs, as these will overpower the taste of the Cacao (e.g. *Artemisia absinthium, Gentiana lutea, Acorus calamus*); likewise, certain aromatics may not combine well with the flavour of Cacao individually or in combination, so should be used with a light hand, or skilfully blended e.g *Salvia officinalis, Thymus vulgaris, Inula helenium*.

A few examples/ideas for herbs to add/combine:

- Erectile dysfunction: *Turnera diffusa, Verbena officinalis, Tilia x europaea, Capsicum minimum, Cordyceps, Lepidium meyenii, Cinnamomum zeylanicum*, possibly *Eryngium maritimum* rad.
- Functional heart issues: *Citrus aurantium, Crataegus* spp., *Leonurus cardiaca, Magnolia mexicana*.
- Hypertension: *Olea x europaea, Achillea millefolium, Tilia* spp., *Valeriana officinalis, Crataegus* spp.
- Lung tonic e.g. chronic inflammatory lung conditions (post-viral syndromes, asthma, cystic fibrosis): add small amounts of mucolytic expectorants (*Hedera helix, Drimia maritima, Primula veris* rad.), aromatic expectorants/anti-infectives (*Hyssopus officinalis, Thymus vulgaris, Inula helenium*), demulcents (*Verbascum thapsus, Althaea officinalis*) and medicinal honeys such as Borage or Manuka.
- Immunomodulatory: add powdered *Trametes versocolor, Ganoderma lucidum* spore extract (much less bitter than the fruiting body), *Astragalus membranaceus, Calendula officinalis* flowers (powder or decoction).
- Nootropic: *Melissa officinalis, Centella asiatica, Salvia rosmarinus, Bacopa monnieri*.

REFERENCES

Chapter One

1. McNeil, in McNeil, 2006a [Book].
2. Beckett, 2000 [Book].
3. Beckett, 2000 [Book].
4. Montamayor *et al.*, 2008 [Journal].
5. Callebaut cholates web article, 2023 [Internet].
6. Montamayor *et al.*, 2008 [Journal].
7. Cornejo *et al.*, 2017 [Internet].
8. Elwers *et al.*, 2009 [Journal].
9. Dreiss & Greenhill, 2008 [Book].
10. Zarillo *et al.*, 2018 [Journal].
11. *Ibid*.
12. Interview with Cacao farmer and APROCAV technical advisor Don Juan Francisco, Cohoban, Alta Verapaz, North Guatemala, 29th January 2011 [see The Secret Life of Chocolate, Appendix C, Interview 16].
13. Young, 1994 [Book].
14. Young, 1994 [Book].
15. Coe, 1994 [Book].
16. Coe & Coe, 1996 [Book].
17. Coe, 1994 [Book].
18. Coe & Coe, 1996 [Book].

19. Coe, 1994 [Book].
20. *Ibid.*
21. Coe & Coe, 1996 [Book].
22. Coe, 1994 [Book].
23. Sahagún, 1950-82, 9:48; *quoted in* Coe, 1996 [Book].
24. Young, 1994 [Book].
25. Sampeck, 2016 [Journal].
26. Young & Severson, 1994 [Journal].

Chapter Two

27. De Montellano, 2004 [Periodical].
28. Miller & Taube, 1993 [Book].
29. Freidel, Schele & Parker, 1993 [Book].
30. Coe & Coe, 1996 [Book].
31. León-Portilla, 1992 [Book].
32. Dreiss & Greenhill, 2008 [Book].
33. Steinbrenner, in McNeil, 2006. [Book].
34. Dreiss & Greenhill, 2008 [Book]; Wilson, in Paoletti *et al.*, 2012 [Book].
35. Rudgley, 1993 [Book].
36. Dreiss & Greenhill, 2008 [Book].
37. *Ibid.*
38. Grivetti, in Grivetti & Shapiro, 2009 [Book].
39. *Ibid.*
40. Eggleston & White, *from* Watson *et al.*, 2013 [Book].
41. 4 February 2011: Interview with midwife Doña Juana Ca'al, Rex'hua, Alta Verapaz, Guatemala; also taken from field notes in conversation with Señora Aurelia Pop, Lanquin, Peten, Guatemala, in 2011.
42. *Ibid.*
43. *Ibid.*
44. Hughes, 1672 [Book].
45. *Ibid.*
46. Awad *et al.* 2003 & 1996 [Journals]; Bouic *et al.*, 1999 [Journal].

47. Colombo, Pinori-Godly & Conti, in Paoletti *et al.*, 2012 [Book].
48. Koga *et al.*, 2022 [Journal].
49. Sudano *et al.*, in Paoletti *et al.*, 2012 [Book].
50. Buitrago-Lopez *et al.*, 2011 [Journal].
51. Zhao *et al.*, 2022 [Journal].
52. Matsumura *et al.*, 2014 [Journal].
53. Grassi *et al.*, 2008 & 2005 [Journals]; Muniyappa *et al.*, 2008 [Journal]; Davison *et al.*, 2008 [Journal]; Faridi *et al.*, 2008 [Journal].
54. Mellor & Nuamovski, 2016 [Journal].
55. Wirtz *et al.*, 2014 [Journal].
56. Martin *et al.*, 2009 [Journal].
57. Buijsse *et al.*, 2006 [Journal].
58. Zhong *et al.*, 2021 [Journal].
59. Eggleston & White, from Watson *et al.*, 2013 [Book].
60. Bisson *et al.*, 2008 [Journal].
61. Osakabe *et al.*, 2009 [Journal].
62. Goya *et al.*, 2016 [Journal].
63. Ohno *et al.*, 2009 [Journal]; Yamagishi *et al.*, 2002 [Journal].
64. Ramiro-Puig & Castell, 2009 [Journal].
65. Neukam *et al.*, 2007 [Journal].
66. Heinrich *et al.*, 2006 [Journal].
67. Williams, Tamburic, & Lally, 2009 [Journals].
68. Extrapolated from Zoumas, Kreiser & Martin, 1980 [Journal].
69. Smit, in Fredholm, 2011 [Book].
70. Baggot *et al.*, 2013 [Journal].
71. Smit, in Fredholm, 2011 [Book].
72. Sugimoto *et al.*, 2014 [Journal].
73. Crichton, Elias, & Alkerwi, 2016 [Journal].
74. Sumiyoshi *et al.*, 2019 [Journal].
75. Eggleston & White, from Watson *et al.*, 2013 [Book].
76. *Ibid.*
77. *Ibid.*
78. Erkkola *et al.*, 2012 [Journal].
79. Becker *et al.*, 2013 [Internet].

80. Stark *et al.*, 2008 [Journal].
81. Usmani *et al.*, 2005 [Journal].
82. Barnes, 2013 [Journal].
83. Simons *et al.*, 1985 [Journal].
84. Girish *et al.*, 2009 [Journal].
85. Wahby *et al.*, 2012 [Journal]; Kemparaju *et al.*, 2006 [Journal].
86. Misztal *et al.*, 2010 [Journal].
87. Quertemont & Didone, 2006 [Journal]; Polache & Granero, 2013 [Journal].
88. Airaksinen & Kari, 1981 [Journal].
89. Subramanian & Prasath, 2014 [Journal]; Zhou, Chan & Zhou, 2012 [Journal].
90. Berry, 2004 [Journal].
91. Marseglia, Palla, & Caligiani, 2014 [Journal].
92. Cornacchia *et al.*, 2012 [Journal].
93. De Laurentiis *et al.*, 2010 [Journal]; Luce *et al.*, 2014 [Journal].
94. Faure & Chapman, 2021 [Internet].
95. di Tomaso, Beltramo & Piomelli, 1996 [Journal]; Beltramo & Piomelli, 1998 [Journal].
96. Urbanska *et al.*, 2019 [Journal].
97. Lin, Wang, Zhou, Xu & Yao, 2022 [Journal].
98. Panneerselvam *et al.*, 2010 [Journal].
99. Sies, Schewe, Heiss & Kelm, 2005 [Journal].
100. Saunders, 1993 [Book].
101. Bluelight.org opiate forum, 2008–2011 [Internet].
102. Parsons, 2021 [Internet].
103. Rusconi *et al.*, in Paoletti *et al.*, 2012 [Book].
104. Rose, Koperski & Golomb, 2010 [Journal].
105. Francis *et al.*, 2006 [Journal].
106. Meier, Noll & Molokwu, 2017 [Journal].
107. Castell *et al.*, in Watson *et al.*, 2013 [Book].
108. Radin, Hayssen, & Walsh, 2007 [Journal].
109. Esch & Stefano, 2005 [Journal].
110. Luce *et al.*, 2014 [Journal].
111. Yamada *et al.*, 2009 [Journal].

112. Räikkönen *et al.*, 2004 [Journal].
113. Strandberg *et al.*, 2008 [Journal].
114. Starr & Starr, in Szogyi, 1997 [Book].

Chapter Three

115. Martin, in McNeil, 2006 [Book].
116. Dreiss & Greenhill, 2008 [Book].
117. Reents-Budet, *from* McNeil, 2006 [Book].
118. Dreiss & Greenhill, 2008 [Book].
119. *Ibid.*
120. *Ibid.*
121. Tedlock, D., 1996 [Book]; Freidel, Schele, & Parker, 1993 [Book].
122. Coe & Coe, 1996 [Book].
123. Grofe, 2007 [Thesis].
124. *Ibid.*
125. *Ibid.*
126. McNeil, in McNeil, 2006b [Book].
127. Miller & Taube, 1993 [Book].
128. Chevalier & Bain, 2002 [Book]; Grofe, 2007 [Thesis].
129. Miller & Taube, 1993 [Book].
130. Dreiss & Greenhill, 2008 [Book].
131. Richter, 2022 [Internet].
132. Grivetti & Cabezon, in Grivetti & Shapiro, 2009 [Book].
133. Martin, in McNeil, 2006 [Book].
134. 26 January 2011: Interview with "Cacao shaman" Keith Wilson, San Marcos La Laguna, Lake Atitlan, Guatemala [Appendix C, Interview 3].
135. Dreiss & Greenhill, 2008 [Book].
136. Coe & Coe, 1996 [Book].
137. Dreiss & Greenhill, 2008 [Book].
138. McNeil, in McNeil, 2006b [Book].
139. Miller & Taube, 1993 [Book].
140. Tedlock, 1992 [Book].
141. Milbrath, 1999 [Book].

142. Freidel, Schele, & Parker, 1993 [Book].

143. Sterger, 2010 [Thesis].

144. Boone, 2007 [Book].

145. Freidel, Schele, & Parker, 1993 [Book].

146. *Ibid.*

147. Faust & López, in McNeil, 2006 [Book].

148. Dreiss & Greenhill, 2008 [Book].

149. Parker & Brotchie, in Paoletti *et al.*, 2012 [Book].

150. Grivetti & Shapiro, 2009 [Book].

151. Ott, 1996 [Book].

152. Mitchell *et al.*, 2021 [Journal].

153. https://www.streetwisdom.org/ (2024) [Internet].

154. Trilling, 1999 [Book]; and one of my clients—see *The Secret Life of Chocolate* Chapter 5 for full anecdote.

155. Rankin *et al.*, 2005 [Journal].

BIBLIOGRAPHY

Books

Beckett, S. (2000). *The Science of Chocolate*. Cambridge: The Royal Society of Chemistry.

Boone, E. (2007). *Cycles of Time and Meaning in the Mexican Books of Fate*. Austin, TX: University of Texas Press.

Castell, M., Pérez-Cano, F., & Bisson, J.-F. (2013). Clinical benefits of cocoa: An overview. In: R. Watson, V. Preedy, & S. Zibadi (Eds.), *Nutrition and Health, Volume 7: Chocolate in Health and Nutrition* (pp. 265–275). New York: Springer, Humana Press.

Chevalier, J., & Bain, A. (2002). *The Hot and the Cold: Ills of Humans and Maize in Native Mexico*. Toronto, Canada: University of Toronto Press.

Coe, M. (2005). *The Maya* [7th edition]. London: Thames & Hudson.

Coe, S. (1994). *America's First Cuisines*. Austin, TX: University of Texas Press.

Colombo, M., Pinorini-Godly, M., & Conti, A. (2012). Botany and pharmacognosy of the Cacao tree. In: R. Paoletti, A. Poli, A. Conti, & F. Visioli (Eds.), *Chocolate and Health* (pp. 41–62). Milan, Italy: Springer-Verlag.

Dreiss, M., & Greenhill, S. (2008). *Chocolate: Pathway to the Gods*. Tucson, AZ: University of Arizona Press.

Eggleston, K., & White, T. (2013). Chocolate and pain tolerance. In: R. Watson, V. Preedy, & S. Zibadi (Eds.), *Nutrition and Health, Volume 7: Chocolate in Health and Nutrition* (pp. 437–447). New York: Springer, Humana Press.

Faust, B., & López, J. (2006). Cacao in the Yukatek Maya healing ceremonies of Don Pedro Ucán Itza. In: C. McNeil (Ed.), *Chocolate in Mesoamerica: A Cultural History of Cacao* (pp. 408–428). Gainesville, FL: University Press of Florida.

Freidel, D., Schele, L., & Parker, J. (1993). *Maya Cosmos: Three Thousand Years on the Shaman's Path*. New York: Harper Collins.

Grivetti, L. (2009). Medicinal chocolate in New Spain, Western Europe, and North America. In: L. Grivetti & H. Shapiro (Eds.), *Chocolate: History, Culture, and Heritage* (pp. 67–88). Hoboken, NJ: John Wiley & Sons.

Hughes, W. (1672). *The American Physitian; or, a treatise of the roots, plants, trees, shrubs, fruit, herbs etc. growing in the English plantations in America […] Whereunto is added a discourse of the cacao-nut-tree, and the use of its fruit; with all the ways of making of chocolate*. London, U.K.: William Cook.

Martin, S. (2006). Cacao in ancient Maya religion: First Fruit from the Maize Tree and other tales from the underworld. In: C. McNeil (Ed.), *Chocolate in Mesoamerica: A Cultural History of Cacao* (pp. 54–183). Gainesville, FL: University Press of Florida.

McNeil, C. (2006a). Introduction. In: C. McNeil (Ed.), *Chocolate in Mesoamerica: A Cultural History of Cacao* (pp. 1–28). Gainesville, USA: University Press of Florida.

Miller, M., & Taube, K. (1993). *An Illustrated Dictionary of The Gods and Symbols of Ancient Mexico and the Maya*. London, U.K.: Thames & Hudson.

Ott, J. (1996). *Pharmacotheon: Entheogenic Drugs, Their Plant Sources and History*. Washington, DC: Natural Products Co.

Parker, G., & Brotchie, H. (2012). Chocolate and mood. In: R. Paoletti, A. Poli, A. Conti, & F. Visioli (Eds.), *Chocolate and Health* (pp. 147–153). Milan, Italy: Springer-Verlag.

Patchett, M. (2020). *The Secret Life of Chocolate*. London, U.K.: Aeon.

Reents-Budet, D., & McNeil, C. (2006). The social context of Kakaw drinking among the ancient Maya. In: C. McNeil (Ed.), *Chocolate in Mesoamerica: A Cultural History of Cacao* (pp. 202–223). Gainesville, FL: University Press of Florida.

Rudgley, R. (1998). *The Encyclopaedia of Psychoactive Substances*. London: Abacus.

Rusconi, M., Rossi, M., Moccetti, T., & Conti, A. (2012). Acute vascular effects of chocolate in healthy human volunteers. In: R. Paoletti, A. Poli, A. Conti, & F. Visioli (Eds.), *Chocolate and Health* (pp. 87–102). Milan, Italy: Springer-Verlag.

Saunders, N. (1993). *E for Ecstasy*. London: Neal's Yard Desktop Publishing Studio.

Smit, H. (2011). Theobromine and the pharmacology of cocoa. In: B. Fredholm (Ed.), *Handbook of Experimental Pharmacology 200* (pp. 201–234). Berlin: Springer-Verlag.

Starr, L., & Starr, E. (1997). Locus of control and chocolate perceptions. In: Szogyi, A. (Ed.), *Chocolate, Food of the Gods* (pp. 11–17). Westford, CT: Greenwood Press.

Sudano, I., Flammer, A., Noll, G., & Corti, R. (2012). Vascular and platelet effects of cocoa. In: R. Paoletti, A. Poli, A. Conti, & F. Visioli (Eds.), *Chocolate and Health*. Milan, Italy: Springer-Verlag.

Tedlock, B. (1992). *Time and the Highland Maya* [Revised Edition]. Albuquerque, NM: University of New Mexico Press.

Tedlock, D. (Trans.) (1996). *Popol Vuh: The Definitive Edition of the Mayan Book of the Dawn of Life and the Glories of Gods and Kings* [2nd Edition]. New York: Touchstone.

Trilling, S. (1999). *Seasons of My Heart: A Culinary Journey Through Oaxaca, Mexico*. New York: Ballantine.

Wilson, P. (2012). Chocolate as medicine: a changing frame-work of evidence throughout history. In: R. Paoletti, A. Poli, A. Conti, & F. Visioli, F. (Eds.), *Chocolate and Health* (pp. 1–16). Milan, Italy: Springer-Verlag.

Young, A. (1994). *The Chocolate Tree*. Washington, DC: Smithsonian Iinstitution Scholarly Press.

Journals & Theses

Airaksinen, M., & Kari, I. (1981). Beta-carbolines, psychoactive compounds in the mammalian body. Part I: Occurrence, origin and metabolism. *Medical Biology, 59*(1): 21–34.

Awad, A., Chen, Y., Fink, C., & Hennessey, T. (1996). Beta-sitosterol inhibits HT-29 human colon cancer cell growth and alters membrane lipids. *Anticancer Research, 16*(5A): 2797–2804.

Awad, A., Roy, R., & Fink, C. (2003). Beta-sitosterol, a plant sterol, induces apoptosis and activates key caspases in MDA-MB-231 human breast cancer cells. *Oncology Reports, 10*(2): 497–500.

Baggott, M., Childs, E., Hart, A., de Bruin, E., Palmer, A., Wilkinson, J., & de Wit, H. (2013). Psychopharmacology of theobromine in healthy volunteers. *Psychopharmacology, 228*(1): 109–118.

Barnes, P. (2013). Theophylline: new perspectives for an old drug. *American Journal of Respiratory and Critical Care Medicine, 167*: 813–818.

Beltramo, M., & Piomelli, D. (1998). Reply: Trick or treat from food endocannabinoids? *Nature, 396*: 636–637.

Berry, M. (2004). Mammalian central nervous system trace amines. Pharmacologic amphetamines, physiologic neuro-modulators. *Journal of Neurochemistry, 90*(2): 257–271.

Bisson, J., Guardia-Llorens, M., Hidalgo, S., Rozan, P., & Messaoudi, M. (2008). Protective effect of Acticoa powder, a cocoa polyphenolic extract, on prostate carcinogenesis in

Wistar-Unilever rats. *European Journal of Cancer Prevention*, *17*(1): 54–61.

Bouic, P., Clark, A., Lamprecht, J., Freestone, M., Pool, E., Liebenberg, R., Kotze, D., & van Jaarsveld, P. (1999). The effects of B-sitosterol (BSS) and B-sitosterol glucoside (BSSG) mixture on selected immune parameters of marathon runners: inhibition of post marathon immune suppression and inflammation. *International Journal of Sports Medicine*, *20*(4): 258–262.

Buijsse, B., Feskens, E., Kok, F., & Kromhout, D. (2006). Cocoa intake, blood pressure, and cardiovascular mortality. The Zutphen Elderly Study. *Archives of Internal Medicine*, *166*(4): 411–417.

Buitrago-Lopez, A., Sanderson, J., Johnson, L., Warnakula, S., Wood, A., Di Angelantonio, E., & Franco, O. (2011). Chocolate consumption and cardiometabolic disorders: systematic review and meta-analysis. *British Medical Journal*, *343*: d4488.

Cornacchia, C., Cacciatore, I., Baldassarre, L., Mollica, A., Feliciani, F., & Pinnen F. (2012). 2,5-diketopiperazines as neuroprotective agents. *Mini Reviews in Medicinal Chemistry*, *12*(1): 2–12.

Crichton, G., Elias, M., & Alkerwi, A. (2016). Chocolate intake is associated with better cognitive function: The Maine-Syracuse Longitudinal Study. *Appetite*, *100*: 126–132.

Davison, K., Coates, M., Buckley, J., & Howe, P. (2008). Effect of cocoa flavanols and exercise on cardiometabolic risk factors in overweight and obese subjects. *International Journal of Obesity*, *32*(8): 1289–1296.

de Laurentiis, A., Fernandez-Solari, J., Mohn, C., Burdet, B., Zorrilla-Zubilete, M., & Rettori, V. (2010). The hypothalamic endocannabinoid system participates in the secretion of oxytocin and tumor necrosis factor-alpha induced by lipopolysaccharide. *Journal of Neuroimmunology*, *221*(1–2): 32–41.

de Montellano, B. (2004). Magia medicinal azteca. *Arqueologia Mexicana*, *12*(69): 30–33.

di Tomaso, E., Beltramo, M., & Piomelli, D. (1996). Scientific correspondence: brain cannabinoids in chocolate. *Nature*, 22(382): 677–678.

Elwers, S., Zambrano, A., Rohsius, C., & Lieberei, R. (2009). Differences between the content of phenolic compounds in Criollo, Forastero and Trinitario cocoa seed (*Theobroma cacao* L.). *European Food Research and Technology*, 229: 937–948.

Erkkola, M., Nwaru, B., Kaila, M., Kronberg-Kippilä, C., Ilonen, J., Simell, O., Veijola, R., Knip, M., & Virtanen, S. (2012). Risk of asthma and allergic outcomes in the offspring in relation to maternal food consumption during pregnancy: a Finnish birth cohort study. *Pediatrica Allergy and Immunology*, 23(2): 186–194.

Esch, T., & Stefano, G. (2005). The Neurobiology of Love. *Neuroendocrinology Letters*, 26(3): 175–190.

Faridi, Z., Njivke, V., Dutta, S., Ali, A., & Katz, D. (2008). Acute dark chocolate and cocoa ingestion and endothelial function: a randomized controlled crossover trial. *American Journal of Clinical Nutrition*, 88(1): 58–56.

Francis, S., Head, K., Morris, P., & Macdonald, I. (2006). The effect of flavanol-rich cocoa on the fMRI response to a cognitive task in healthy young people. *Journal of Cardiovascular Pharmacology*, 47, Suppl. 2: S215–220.

Girish, K., Kemparaju, K., Nagaraju, S., & Vishwanath, B. (2009). Hyaluronidase inhibitors: a biological and therapeutic perspective. *Current Medicinal Chemistry*, 16(18): 2261–2288.

Goya, L., Martín, M., Sarriá, B., Ramos, S., Mateos, R., & Bravo, L. (2016). Effect of cocoa and its flavonoids on biomarkers of inflammation: Studies of cell culture, animals and humans. *Nutrients*, 8(4): 212.

Grassi, D., Lippi, C., Necozione, S., Desideri, G., & Ferri, C. (2005). Short-term administration of dark chocolate is followed by a significant increase in insulin sensitivity and a decrease in blood pressure in healthy persons. *American Journal of Clinical Nutrition*, 81(3): 611–614

Grassi, D., Lippi, C., Necozione, S., Desideri, G., & Ferri, C. (2008). Blood pressure is reduced and insulin sensitivity increased in glucose-intolerant, hypertensive subjects after 15 days of consuming high-polyphenol dark chocolate. *Journal of Nutrition*, *139*(9): 1671–1676.

Grofe, M. (2007). The recipe for rebirth: Cacao as fish in the mythology and symbolism of the ancient Maya. [Thesis: Department of Native American Studies, University of California at Davis.] www.famsi.org/research/grofe/GrofeRecipeForRebirth.pdf [Accessed 4 April 2019].

Heinrich, U., Neukam, K., Tronnier, H., Sies, H., & Stahl, W. (2006). Long-term ingestion of high flavanol cocoa provides photoprotection against UV-induced erythema and improves skin condition in women. *Journal of Nutrition*, *136*(6): 1565–1569.

Kemparaju, K., & Girish, K. (2006). Snake venom hyaluronidase: a therapeutic target. *Cell Biochemistry & Function*, *24*(1): 7–12.

Koga, J., Ojiro, K., Yanagida, A., Suto, T., Hiki, H., Inoue, Y., Sakai, C., Nakamoto, K., Fujisawa, Y., Orihara, A., Murakami, H., Hirasawa, S., Nakajima, K., Sakazawa, T., & Yamane, H. (2022). Ingestion of Indigestible Cacao Proteins Promotes Defecation and Alters the Intestinal Microbiota in Mice. *Current Developments in Nutrition*. 6; 6 (10).

Jianguo, L., Wang, Q., Zhou, S., Xu, S., & Yao, K. (2022). Tetramethylpyrazine: A review on its mechanisms and functions. *Biomedicine & Pharmacotherapy*. Jun; 150: 113005.

Luce, V., Fernandez Solari, J., Rettori, V., & De Laurentiis, A. (2014). The inhibitory effect of anandamide on oxytocin and vasopressin secretion from neurohypophysis is mediated by nitric oxide. *Regulatory Peptides*, *188*: 31–39.

Marseglia, A., Palla, G., & Caligiani, A. (2014). Presence and variation of γ-aminobutyric acid and other free amino acids in cocoa beans from different geographical origins. *Food Research International*, *63*, Part C: 360–366.

Martin, F., Rezzi, S., Peré-Trepat, E., Kamlage, B., Collino, S., Leibold, E., Kastler, J., Rein, D., Fay, L., & Kochhar, S. (2009). Metabolic effects of dark chocolate consumption on energy, gut microbiota, and stress-related metabolism in free-living subjects. *Journal of Proteome Research*, *8*(12): 5568–5579.

Matsumura, Y., Nakagawa, Y., Mikome, K., Yamamoto, H., & Osakabe, N. (2014). Enhancement of energy expenditure following a single oral dose of flavan-3-ols associated with an increase in catecholamine secretion. *PLOS One*, *9*(11): e112180.

Meier, B., Noll, S., & Molokwu, O. (2017). The sweet life: The effect of mindful chocolate consumption on mood. *Appetite*, *108*: 21–27.

Mellor, D., & Naumovski, N. (2016). Effect of cocoa in diabetes: the potential of the pancreas and liver as key target organs, more than an antioxidant effect? *International Journal of Food Science and Technology*, *51*(4): 829–841.

Misztal, T., Tomaszewska-Zaremba, D., Górski, K., & Romanowicz, K. (2010). Opioid-salsolinol relationship in the control of prolactin release during lactation. *Neuroscience*, *170*(4): 1165–1171.

Mitchell, J., Bogenschutz, M., Lilienstein, A., Harrison, C., Kleiman, S., Parker-Guilbert, K., Ot'alora, M., Garas, W., Paleos, C., Gorman, I., Nicholas, C., Mithoefer, M., Carlin, S., Poulter, B., Mithoefer, A., Quevedo, S., Wells, G., Klaire, S., van der Kolk, B., Tzarfaty, K., Amiaz, R., Worthy, R., Shannon, S., Woolley, J., Marta, C., Gelfand, Y., Hapke, E., Amar, S., Wallach, Y., Brown, R., Hamilton, S., Wang, J., Coker, A., Matthews, R., de Boer, A., Yazar-Klosinski, B., Emerson, A., & Doblin, R. (2021). MDMA-assisted therapy for severe PTSD: a randomized, double-blind, placebo-controlled phase 3 study. *Nature Medicine*, *27*(6), 1025–1033.

Montomayor, J., Lachenaud, P., da Silva e Mota, J., Loor, R., Kuhn, D., Brown, J., & Schnell, R. (2008). Geographic and genetic population differentiation of the Amazonian

chocolate tree. *PLoS One, 3*(10): e3311. doi: 10.1371/journal.
pone.0003311.

Muniyappa, R., Hall, G., Kolodziej, T., Karne, R., Crandon, S.,
& Quon, M. (2008). Cocoa consumption for 2 wk enhances
insulin-mediated vasodilatation without improving blood
pressure or insulin resistance in essential hypertension.
American Journal of Clinical Nutrition, 88(6): 1685–1696.

Neukam, K., Stahl, W., Tronnier, H., Sies, H., & Heinrich, U.
(2007). Consumption of flavanol-rich cocoa acutely increases
microcirculation in human skin. *European Journal of Nutrition, 46*(1): 53–56.

Ohno, M., Sakamoto, K., Ishizuka, K., & Fujita, S. (2009). Crude
cacao *theobroma cacao* extract reduces mutagenicity induced
by benzo[a]pyrene through inhibition of CYP1A activity *in
vitro. Phytotherapy Research, 23*(8): 1134–1139.

Osakabe, N., Baba, S., Yasuda, A., Iwamoto, T., Kamiyama, M.,
Takizawa, T., Itakura H., & Kondo, K. (2009). Daily cocoa
intake reduces the susceptibility of low-density lipoprotein
to oxidation as demonstrated in healthy human volunteers.
Free Radical Research, 34(1): 93–99.

Panneerselvam, M., Tsutsumi, Y., Bonds, J., Horikawa, Y.,
Saldana, M., Dalton, N., Head, B., Patel, P., Roth, D., &
Patel, H. (2010). Dark chocolate receptors: epicatechin-
induced cardiac protection is dependent on δ-opioid recep-
tor stimulation. *American Journal of Physiology—Heart and
Circulatory, 299*(5): H1604–H1609.

Polache, A., & Granero, L. (2013). Salsolinol and ethanol-derived
excitation of dopamine mesolimbic neurons: new insights.
Frontiers in Behavioural Neuroscience, 7, Article 74: 1–2.

Quertemont, E., & Didone, V. (2006). Role of acetaldehyde in
mediating the pharmacological and behavioral effects of
alcohol. *Alcohol Research and Health, 29*(4): 258–265.

Radin, D., Hayssen, G., & Walsh, J. (2007). Effects of intentionally enhanced chocolate on mood. *Explore: The Journal of Science and Healing.* Vol. 3, No. 5, 485–492.

Räikkönen, K., Pesonen, A., Järvenpää, A., & Strandberg, T. (2004). Sweet babies: chocolate consumption during pregnancy and infant temperament at six months. *Early Human Development*, 76(2): 139–145.

Ramiro-Puig, E., Casadesús, G., Lee, H., Zhu, X., McShea, A., Perry, G., Pérez-Cano, F., Smith, M., & Castell, M. (2009). Neuroprotective effect of cocoa flavonoids on in vitro oxidative stress. *European Journal of Nutrition*, 48(1): 54–61.

Rose, N., Koperski, S., & Golomb, B. (2010). Mood food: chocolate and depressive symptoms in a cross-sectional analysis. *Archives of Internal Medicine*, 170(8): 699–703.

Sampeck, K. (2016). Cacao biology: chocolate culture, a superfood. *ReVista: Harvard Review of Latin America*, 16(1): 3–9.

Sies, H., Schewe, T., Heiss, C., & Kelm, M. (2005). Cocoa polyphenols and inflammatory mediators. *American Journal of Clinical Nutrition*. 81 (1 Suppl.) 304S–312S.

Simons, E., Becker, A., Simons, K., & Gillespie, C. (1985). The bronchodilator effect and pharmacokinetics of theobromine in young patients with asthma. *Journal of Allergy and Clinical Immunology*, 76(5): 703–707.

Stark, T., Lang, R., Keller, D., Hensel, A., & Hofmann, T. (2008). Absorption of N-phenylpropenoyl-L-amino acids in healthy humans by oral administration of cocoa (Theobroma cacao). *Molecular Nutrition and Food Research*, 52(10): 1201–1214.

Sterger, K. (2010). Crosses, flowers and toads: Maya bloodletting iconography in Yaxchilan lintels 24, 25 and 26. [Thesis—Department of Humanities, Classics, and Comparative Literature, Brigham Young University.] https://scribd.com/document/71503199/Crosses-Flowers-And-Toads-Maya-Blodletting-Iconography [Accessed 4 April 2019].

Strandberg, T., Strandberg, A., Pitkälä, K., Salomaa, V., Tilvis, R., & Miettinen, T. (2008). Chocolate, well-being and health among elderly men. *European Journal of Clinical Nutrition*, 62(2): 247–253.

Subramanian, S., & Prasath, G. (2014). Antidiabetic and anti-dyslipidemic nature of trigonelline, a major alkaloid of fenugreek seeds studied in high-fat-fed and low-dose strep-tozotocin-induced experimental diabetic rats. *Biomedicine & Preventive Nutrition.* Volume 4, Issue 4, 475–480.

Sugimoto, N., Miwa, S., Hitomi, Y., Nakamura, H., Tsuchiya, H., & Yachie, A. (2014). Theobromine, the primary methylx-anthine found in Theobroma cacao, prevents malignant glioblastoma proliferation by negatively regulating phos-phodiesterase-4, extracellular signal-regulated kinase, Akt/mammalian target of rapamycin kinase, and nuclear factor-kappa B. *Nutrition and Cancer, 66*(3): 419–423.

Sumiyoshi, E., Matsuzaki, K., Sugimoto, N., Tanabe, Y., Hara, T., Katakura, M., Miyamoto, M., Mishima, S., & Shido, O. (2019). Sub-Chronic Consumption of Dark Choc-olate Enhances Cognitive Function and Releases Nerve Growth Factors: A Parallel-Group Randomized Trial. *Nutri-ents.* 11, (11), 2800–2815.

Urbańska, B., Derewiaka, D., Lenart, A., & Kowalska, J. (2019). Changes in the composition and content of polyphenols in chocolate resulting from pre-treatment method of cocoa beans and technological process. *European Food Research and Technology.* Vol. 245, 2101–2112.

Usmani, O., Belvisi, M., Patel, H., Crispino, N., Birrell, M., Korbonits, M., Korbonits, D., & Barnes, P. (2005). Theobro-mine inhibits sensory nerve activation and cough. *FASEB Journal, 19*(2): 231–233.

Wahby, A., Mahdy, E., El-Mezayn, H., Salama, W., Ebrahim, N., Abdel-Aty, A., & Fahmy, A. (2012). Role of hyaluronidase inhibitors in the neutralization of toxicity of Egyptian horned viper *Cerastes cerastes* venom. *Journal of Genetic Engi-neering and Biotechnology, 10*(2): 213–219.

Williams, S., Tamburic, S., & Lally, C. (2009). Eating chocolate can significantly protect the skin from UV light. *Letters in Applied Microbiology, 49*(3): 354–360.

Wirtz, P., Känel, R., Meister, R., Arpagaus, A., Treichler, S., Kuebler, U., Huberk, S., & Ehlert, U. (2014). Research correspondence: Dark chocolate intake buffers stress reactivity in humans. *Journal of the American College of Cardiology, 63*(21): 2297–2299.

Yamada, T., Yamada, Y., Okano, Y., Terashima, T., & Yokogoshi, H. (2009). Anxiolytic effects of short- and long-term administration of cacao mass on rat elevated T-maze test. *Journal of Nutritional Biochemistry, 20*(12): 948–955.

Yamagishi, M., Natsume, M., Osakabe, N., Nakamura, H., Furukawa, F., Imazawa, T., Nishikawa, A., & Hirose, M. (2002). Effects of cacao liquor proanthocyanidins on PhIP-induced mutagenesis in vitro, and in vivo mammary and pancreatic tumorigenesis in female Sprague–Dawley rats. *Cancer Letters, 185*(2): 123–130.

Young, A., & Severson, D. (1994). Comparative analysis of steam distilled floral oils of cacao cultivars (*Theobroma cacao* L., Sterculiaceae) and attraction of flying insects: Implications for a *Theobroma* pollination syndrome. *Journal of Chemical Ecology, 20*(10): 2687–2703.

Zarrillo, S., Gaikwad, N., Lanaud, C., Powis, T., Viot, C., Lesur, I., Fouet, O., Argout, X., Guichoux, E., Salin, F., Solorzano, R., Bouchez, O., Vignes, H., Severts, P., Hurtado, J., Yepez, A., Grivetti, L., Blake, M., & Valdez, F. (2018). The use and domestication of *Theobroma cacao* during the mid-Holocene in the upper Amazon. *Nature Ecology & Evolution*. Volume 2, 1879–1888.

Zhao, B., Gan, L., Yu, K., Männistö , Huang, J., & Albanes, D. (2022). Relationship between chocolate consumption and overall and cause-specific mortality, systematic review and updated meta-analysis. *European Journal of Epidemiology*. 37, (4), 321–333.

Zhong , G., Hu, T., Yang, P., Peng, Y., Wu, J., Sun, W., Cheng, L., & Wang, C. (2021). Chocolate consumption and all-cause and cause-specific mortality in a US population: a post hoc analysis of the PLCO cancer screening trial. *Aging (Albany NY)*. 13, (14), 18564–18585.

Zhou, J., Chan, L., & Zhou, S. (2012). Trigonelline: a plant alkaloid with therapeutic potential for diabetes and central nervous system disease. *Current Medicinal Chemistry*. 19, (21), 3523–31.

Zoumas, B., Kreiser, W., & Martin, R. (1980). Theobromine and caffeine content of chocolate products. *Journal of Food Science*, *45*(2): 314–316.

Web Pages

Becker, K., Gesiler, S., Ueberall, F., Fuchs, D., & Gostner, J. (2013). Immunomodulatory properties of cacao extracts—potential consequences for medical applications. *Frontiers in Pharmacology*. https://doi.org/10.3389/fphar.2013.00154 [Accessed 4 March 2019].

Cornejo, O., Muh-Ching, Y., Dominguez, V., Andrews, M., Sockell, A., Møller, E., Livingstone, D., Stack, C., Romero, A., Umaharan, P., Royaert, S., Tawari, N., Ng, P., Schnell, R., Phillips, W., Mockaitis, K., Bustamante, C., & Motamayor, J. (2017). Genomic insights into the domestication of the chocolate tree, *Theobroma cacao* L. *bioRxiv: The Preprint Server for Biology*. https://doi.org/10.1101/223438 [Accessed May 2023].

Faure, L., & Chapman, K. (2021). N-Acylphosphatidylethanolamines (NAPEs), N-acylethanolamines (NAEs) and Other Acylamides: Metabolism, Occurrence and Functions in Plants. *AOCS Lipid Library: The American Oil Chemists' Society*. https://lipidlibrary.aocs.org/chemistry/physics/plant-lipid/n-acylphosphatidylethanolamines-(napes)-n-acylethanolamines-(naes)-and-other-acylamides-metabolism-occurrence-and-functions-in-plants [Accessed May 2023].

Harper, D. (n.d.). Etymology of pharmacology. *Online Etymology Dictionary*. Retrieved January 23, 2023, from https://www.etymonline.com/word/pharmacology

Parsons, A. (2021). The Aztec Combo: Magic Mushrooms & Cacao. *Zamnesia*. https://www.zamnesia.com/blog-aztec-magic-mushrooms-cacao-n383 [Accessed May 2023].

Richter, F. (2022) (Not) Everybody Loves Chocolate. July 7, https://www.statista.com/chart/3668/the-worlds-biggest-chocolate-consumers/ [Accessed 21 April 2023].

(2008). Forum post by "Papa Verine", 8 July. http://forum.opiophile.org/archive/index.php/t-17550.html [Accessed c. 2014].

(2011). Forum post by "ambigroove", 16 December, http://bluelight.org/vb/threads/602275-Chocolate-Opiate-Potentiator [Accessed c. 2014].

(2011). Forum post by "Znegative", 16 December, http://bluelight.org/vb/threads/602275-Chocolate-Opiate-Potentiator [Accessed c. 2014].

(2023). Theobroma cacao, the food of the gods. *Barry Callebaut*. https://www.barry-callebaut.com/about-us/media/press-kit/history-chocolate/theobroma-cacao-food-gods [Accessed 16 August 2018].

(2024). Street Wisdom. https://www.streetwisdom.org/ [Accessed 26 January 2024].

ACKNOWLEDGEMENTS

All my interviewees, for their time and generosity: Reginaldo Chayax Huex of Asociacion Bio-Itza, San José, Peten; Carina Santiago and her student, Kalisa Wells; Keith Wilson; Doña Rosa Gregorio and Roberto Gregorio; Señora Anna Maria Garcia Vasquez and Roberto Vasquez; Mario Euan; Juan Francisco Tzir; Ignacio Cac Sacul; Don Mateo Poptchu; Don Angel Chiac; Doña Juana Ca'al; Don Antonio Xoc; Señora Teresa Olivera; Tomas Villanueva; Doña Rosalia of Lanquin; Señora Lopreto and her daughter; Doña Maria Eugenia Navarrete Matei and her mother; Señora Aurelia Pop; Señora Delfina Valverde; Señora Rosia Torres Torres and Jorge Perez Leon; Señora Claudia Maribel Ya'teni; Don Ramiro; Señora Ernestina Tawal; José Antonio; Claudia at Hacienda la Luz in Tabasco.

Juan-Pablo Porres Esquina, and Alejandra Martinez Porres Esquina; Susana Trilling; Verónica Caal; Maribel Omaña Mendez and Arnaud; Lupita Juarez; Roberto Molina; Veronica Juarez; Emmanuel Cruz Canela; Crispina Navarro Gomez and family; Don Placido Castiliano; Mario Gonzalez; Rocio Remedios; Alfredo Ramirez Rosario; Esteban and his mother at Hotel Don Juan Metalbatz in Coban; Casa de los Amigos, Mexico City. All these people helped to make my travels fruitful and safe.

James and Jennifer Maloney; Heidi Nygård; Omar Ramirez Casas; Bridget Price, Victoria Masters, and Victoria Gervasi for connections and suggestions; Vincent James; Kiralee; Rae; Maya Adar; Julissa in San Antonio Suchitequepez, for the fried breakfasts; Daniela, the accidental American in San Luís, Peten; Eric Pecay; Erdi Debesai; Luciana Morera; Dr Celia Bell of Middlesex University, for advice and support; and Chantal Coady of Rococo Chocolates.

Author inspiration: Jonathan Ott, for his book on chocolate.

Olly and all at Aeon; and all my wonderful former students, colleagues, and clients who have been supportive and interested.

INDEX